Molecular Spectroscopy
With Neutrons

Molecular Spectroscopy With Neutrons

Henri Boutin
Sidney Yip

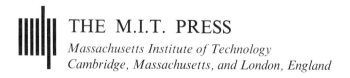

THE M.I.T. PRESS

Massachusetts Institute of Technology
Cambridge, Massachusetts, and London, England

Copyright © 1968 by
The Massachusetts Institute of Technology

Set in Monophoto Times Roman.
Printed and bound in the United States of America by
The Maple Press Company, York, Pennsylvania.

Library of Congress catalog card number: 68–22823

To Phoebe and Nita

Preface

The quantitative study of the structure and properties of matter necessarily involves detailed considerations of the forces acting between atoms and molecules. The determination of frequencies of vibrations between atoms in a molecule (internal modes) and between molecules in a solid or liquid (external modes) is a basic problem in molecular spectroscopy, and a knowledge of these frequencies leads directly to information about interatomic force constants and inter-molecular potential barrier heights. Such investigations are traditionally carried out by means of optical measurements such as infrared absorption and Raman scattering. These are powerful techniques which have been highly refined, and their use has resulted in extensive and valuable data.

Recently the availability of large fluxes of low-energy neutrons from nuclear reactors has enabled new investigations of molecular motions to be initiated. The technique of inelastic neutron scattering, while in a state of relative infancy, has already produced useful new information about crystal dynamics and atomic motions in simple liquids. There seems to be little doubt that such studies will grow in scope and importance. The application of the neutron method to molecular spectroscopy has received less attention; however, even here there exists sufficient evidence to show that research in this direction is potentially very fruitful.

The primary purpose of this work is to describe the use of neutron scattering in the study of molecular solids and liquids. We have addressed our discussions to those who work with neutrons as a tool in molecular research as well as to those spectroscopists and workers in related fields who may be interested in the results which the technique can provide. We have attempted to relate our work to optical spectroscopy. In particular, we have examined in general theoretical terms the connection between neutron-scattering and infrared-absorption experiments. Throughout the book the complementary nature of the two types of data is emphasized.

The work described in this monograph is based on the results of a number of inelastic-neutron-scattering experiments which we have carried out with various collaborators over the past six years. These results, most of which have been reported in journals and at meetings, are assembled here in an attempt to give a unified and systematic treatment of the use of neutrons in molecular spectroscopy. It is hoped that the discussions will make the method more accessible to spectroscopists in general, and that these workers will actively participate in future research efforts using the neutron technique.

Although neutron experiments require facilities not normally found in molecular spectroscopy laboratories, it should be pointed out that these investigations need not be confined to programs feasible only at large nuclear research centers. Most of the measurements discussed in the following, with a few exceptions, were performed with a small nuclear reactor (one megawatt power, peak neutron flux $> 10^{13}$ neutrons/cm^2/sec) and relatively unsophisticated electronic and time-of-flight equipment. A brief description of the various parts of the neutron spectrometer and their assembly is included in this monograph.

Wherever comparable facilities can be found it would seem that a program of molecular research with neutrons would be particularly appropriate. It offers prospects of original research in the area of spectroscopy concerned with molecular motions of very low frequency (0–1000 cm^{-1}). This includes the investigation of intermolecular motions and lattice modes (in crystalline solids), which has received little attention up to now for lack of appropriate experimental tools. The neutron-scattering measurements are complementary to the far infrared measurements, and their importance will grow because these low-frequency molecular motions are necessary to the understanding of the transport properties of liquids and solids. Molecular research with neutrons offers the additional advantage in universities or research establishments of bringing into close collaboration scientists of various disciplines such as physics, chemistry, and biology. In many developing countries,* only a broad research program encompassing many disciplines can provide the type of stimulating environment necessary to attract and keep their own scientists. For those countries which already possess modest nuclear reactor facilities it would seem that a program in materials research with neutrons is an appropriate scientific endeavor. Because of its practical significance, it is likely to find interest and support at the governmental level. Because of its basic nature, it also creates a strong motivation for cooperation among nations through individual scientists, academic and research institutions, and atomic energy agencies.

Cambridge, Mass.   H. BOUTIN and S. YIP
8 January 1968

---

*Countries such as Indonesia, Thailand, Korea, S. Vietnam, Mexico, Brazil, Phillipines, Turkey, Greece, Republic of the Congo, to name only a few, have acquired nuclear reactors with the assistance of the United States.

Acknowledgments

We have benefited a great deal from discussions with A. Agrawal, S. H. Chen, R. C. Desai, R. G. Gordon, J. S. King, P. C. Martin, M. Nelkin, I. Oppenheim, R. K. Osborn, K. Otnes, H. Prask, J. J. Rush, S. Trevino, G. Venkataraman, and T. Wall. In one way or another they have helped us in shaping our own thinking in this field of research. Many of the experiments described here were carried out by one of us (H.B.) in collaboration with H. Prask, S. Trevino, G. Safford, H. Danner, V. D. Gupta, and W. Whittemore over a period of five years. One of us (S.Y.) expresses his appreciation for the financial support given by the Explosives Research Laboratory of Picatinny Arsenal. We are grateful to M. Benedict, J. V. R. Kaufman, J. J. O'Connor, and H. Priest for the support and encouragement they have given us during the course of this work. We are deeply indebted to Mrs. Lenore Stevens for her skill, patience, and care in correcting and editing the manuscript, and we would also like to thank Miss Cheryl Ferger for her assistance in the preparation of the manuscript.

The following illustrations were reproduced with permission.

Figures 3.1 and 3.2 from R. C. Desai and S. Yip, *Phys. Rev.* **166,** 129 (1968).
Figure 3.9 from A. K. Agrawal and S. Yip, *J. Chem. Phys,* **46,** 1999 (1967).
Figure 4.3 from S. Trevino, *J. Chem. Phys.* **45,** 757 (1966).

Figure 4.7 from S. Trevino and H. Boutin, *J. Chem. Phys.* **45,** 2700 (1966).

Figures 5.2–5.4 from V. D. Gupta, S. Trevino, and H. Boutin, *J. Chem. Phys.* **48,** 3008 (1968).

Figures 6.1–6.3 from H. Boutin, G. J. Safford, and V. Brajovic, *J. Chem. Phys.* **39,** 3135 (1963).

Figure 6.5 from H. Boutin and G. J. Safford, in *Inelastic Scattering of Neutrons in Solids and Liquids* (International Atomic Energy Agency, Vienna, 1965), Vol. 2, p. 393.

Figure 6.6a from Y. Imry, I. Pelah, and E. Weiner, *J. Chem. Phys.* **43,** 2332 (1965).

Figure 7.1 from G. Safford, V. Brajovic, and H. Boutin, *J. Phys. Chem. Solids* **24,** 771 (1963). Reprinted with permission from Pergamon Press, New York.

Figure 7.2 from I. Pelah, K. Krebs, and Y. Imry, *J. Chem. Phys.* **43,** 1864 (1965).

Figure 8.1 from J. J. Rush, T. I. Taylor, and W. W. Havens, Jr., *J. Chem. Phys.* **37,** 234 (1962).

Figure 8.2 from G. Venkataraman, K. U. Deniz, P. K. Iyengar, P. R. Vijayaraghavan, and A. P. Roy, *Solid State Commun.* **2,** 17 (1964). Reprinted with permission from Pergamon Press, New York.

Figure 8.3 from H. Boutin, S. Trevino, and H. Prask, *J. Chem. Phys.* **45,** 401 (1966).

Figure 8.4 from V. Brajovic, H. Boutin, G. J. Safford, and H. Palevsky, *J. Phys. Chem. Solids* **24,** 617 (1963).

Figures 9.1–9.6 from H. Prask, H. Boutin, and S. Yip, *J. Chem. Phys.* **48,** 3367 (1968).

Figures 9.7, 9.9, 9.10, 10.1, 10.4, 10.6–10.8, 10.11, and 12.1 from H. Boutin, H. Prask, and R. D. Iyengar, *Advan. Colloid Interface Sci.* **2,** 1 (1966).

Figure 9.8 from D. J. Hughes, H. Palevski, W. Kley, and E. Tunkelo, *Phys. Rev.* **119,** 878(1960).

Figure 9.12 from H. Prask and H. Boutin, *J. Chem. Phys.* **45,** 699 (1966).

Figures 9.13 (lower part) and 10.10 from H. Boutin, G. J. Safford, and H. R. Danner, *J. Chem. Phys.* **40,** 2670 (1964).

Figure 9.14 from K. E. Larsson and U. Dahlberg, *Physica* **30,** 1557 (1964).

Contents

Part I
Introduction

1. General Remarks

Neutrons of very low energy (thermal or subthermal) are a unique probe for the study of molecular dynamics in solids and liquids. Among the various molecular modes, the center-of-mass translation and rotation are of particular interest because of their more sensitive dependence upon the intermolecular forces and the physical state of the medium. Since the neutron energy is comparable to lattice vibrations and molecular rotations, these motions can be observed with relative ease and studied in detail by means of inelastic scattering. Another characteristic of the probe is that the neutron mass is of the same order as the mass of the scattering nuclei. This means that the de Broglie wavelength associated with the momentum transfer is comparable to interatomic distances in condensed matter, and the scattering is therefore sensitive to the structure of the system. For these reasons inelastic neutron scattering is a powerful tool for studying the microscopic environment in solids and liquids. On the other hand, because neutron spectra contain information about motion and structure they are usually quite complex and require considerable analysis in their interpretation. While the complexity of the spectra does not diminish the potential usefulness of neutron studies, it does impose on these investigations a strong dependence on theoretical calculation.

In an inelastic-scattering experiment, the variation of scattering intensity with neutron energy and momentum transfer is observed. The energy and momentum transfers may be expressed as

$$\hbar\omega = E_f - E_i = \frac{\hbar}{2m}(k_f^2 - k_i^2),$$

$$\hbar\,\kappa = \hbar\,(\mathbf{k}_i - \mathbf{k}_f),$$

where E, $\mathbf{k}$, and m denote the neutron energy, wave vector, and mass, and where subscripts i and f denote initial and final conditions. Aside from well-known factors, the cross-section dependence on dynamics and structure is entirely specified by a function only of ω and κ. These are therefore the appropriate variables in which to discuss general cross-section behavior. Moreover, if we simply regard ω and κ as arbitrary frequency and wave vector, we can characterize neutron scattering by the range of ω and κ in which measurements are carried out. In this way, one can compare the neutron probe with other spectroscopic techniques as shown in Figure 1.1. The comparison should not

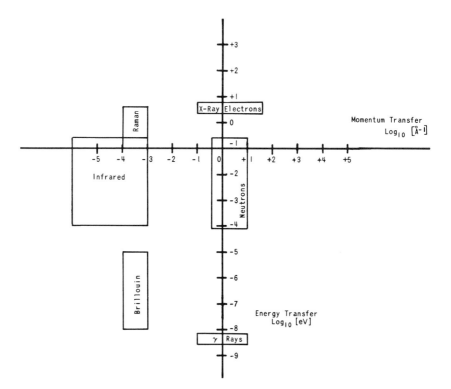

Figure 1.1. Comparison of neutron scattering with other techniques.

be taken too literally since we have not considered the precise nature of the interaction between various radiations and the medium. It is interesting to note that for molecular phenomena the relevant energies are of the order of 10^{-3} to 10^{-1} eV and the wavelengths are of the order of 10^{-8} cm. Thus neutron scattering is presently the only probe capable of revealing the details of dynamical processes on such short space and time scales. In the case of optical scattering and absorption, the frequency range is essentially the same, but the wavelengths are longer by three or more orders of magnitude.

Neutrons can "probe" a molecular system on a time scale between 10^{-14} and 10^{-11} sec. Thus both vibrational motions (of the order of 10^{13} cps) and diffusive motions can be observed. The time scales for observing diffusion are short enough that "microscopic" diffusive processes are observed rather than an average behavior. It is interesting to compare this time scale with the characteristic time scales of the NMR (nuclear magnetic resonance) technique (about 10^{-9} sec) or the even longer time scales involved in dielectric and ultrasonic absorptions. In the neutron measurements, the vibrational motions are observed in a frequency range (30–1000 cm^{-1}) which contains, in general, the intramolecular and all the intermolecular vibrations.

The interaction of neutrons with nuclei is unique for still another reason. The neutron-nucleus interaction is usually approximated by a pseudopotential characterized by a scattering length. The scattering length is a property of the nucleus and therefore can be different for different isotopes and can vary with the spin of the target nucleus. The presence of an isotopic mixture in the scattering system or of spin-dependent interactions gives rise to a process known as incoherent scattering. This type of scattering is basically different from the normal coherent scattering, which is analogous to the scattering of electromagnetic radiation. The scattering of a neutron wave from a system of identical nuclei having zero spin would be, except for magnetic interactions with the atomic electrons, entirely coherent. For such processes, interference among the scattered waves is possible. For a system of identical nuclei with nonzero spin, the neutron-nucleus interaction depends on the relative orientations of the neutron and nuclear spins and therefore gives rise to incoherence. In this case each nucleus scatters independently and no interference effects are observed. For nuclei with zero spin, incoherence can still arise if the scattering length can fluctuate from nucleus to nucleus because of isotopic effects.

Together, coherent and incoherent scattering allow a complete investigation of the structure and motions in the scattering system. The relative amounts of incoherent and coherent contribution present in an experiment, crudely speaking, can be estimated by first considering the nuclear scattering lengths of the sample. This is useful because different calculations are required for the two types of processes. All the molecular systems with which we are primarily concerned contain an appreciable amount of hydrogen. Because hydrogen scattering is essentially incoherent and the cross section is at least an order of magnitude greater than that for other elements, our discussions therefore will deal mainly with the principles and applications of neutron incoherent scattering. This restriction is not an inherent limitation of the neutron technique; rather it is a reflection of our own interest and experience and, to a large extent, of the current state of the field.

The neutron-scattering cross section measured in an experiment generally consists of two parts, elastic and inelastic. The inelastic part contains information about the vibrational motions of the atoms or molecules which can be obtained from the peak positions, their intensity, and their width. Such peaks correspond to a maximum in the density of vibrational states $g(\omega)$ of the system, which are populated at a given temperature according to a Boltzman distribution. The function $g(\omega)$, which expresses the density of vibrational frequencies, is related to the partition function of the system and therefore allows the determination of the thermodynamic properties. The intensity of a peak in $g(\omega)$ also depends upon the amplitude of vibration of normal modes of the spectrum. The sharpness of a band in the inelastic-scattering region depends on several factors: (1) the dispersion of the corresponding branch of the normal mode (in the case of a crystal); (2) the strength of the bond holding a molecule in the lattice; (3) the amount of long-range ordering in the system; and (4) the degree of anharmonicity of the vibrational mode. (The lifetime of a phonon is inversely proportional to the width of the peak and depends upon the amount of phonon-phonon interaction.)

Elastic scattering, on the other hand, provides information about diffusive motions in the scattering system. These diffusive processes manifest themselves as a broadening of the elastic peak in the scattered-neutron spectrum. In the case of a gas there is no elastic peak because the atom is completely free to recoil. In the case of solids and liquids, where one can measure both the eleastic and the inelastic portions of the incoherent-scattering cross section, the most commonly used techniques are the chopper time-of-flight technique and the rotating-crystal time-of-flight technique. Both techniques involve a pulsed beam of monoenergetic neutrons, the energy distribution of the neutrons scattered through various angles being measured by time-of-flight.

The coherent inelastic scattering of neutrons has been measured for materials that can be obtained in single-crystal form (for example, metals, semiconductors). Using a three-axis crystal spectrometer, energy distribution measurements at constant momentum transfer (the "constant-Q" method) can be performed. The resulting dispersion curves of the lattice phonons may in turn be compared with those calculated from specific crystal models. A knowledge of dispersion relations is essential to an understanding of the microscopic properties of materials, and neutron scattering is the most effective tool available for such measurements at the present time.

Throughout the entire monograph we shall emphasize the use of frequency distribution functions in the analysis of incoherent neutron spectra. This approach has as its basis a well-known result for monatomic cubic crystals and the formal extensions to monatomic solids and liquids in general. The frequency distribution function most familiar in this connection is that of the normal modes of lattice vibrations in a harmonic crystal. For a lattice of N unit cells, each with n particles, there are $3n$ normal-mode frequencies (three acoustic branches and the remaining optic branches) for each discrete value of the wave vector in the first Brillouin zone. The frequency distribution $g(\omega)$ can be computed if the dispersion (frequency–wave number) relation is known. However, a knowledge of $g(\omega)$ is generally not sufficient to analyze neutron spectra of molecular systems because, except in the special case of monatomic cubic crystals, the cross section is not simply related to $g(\omega)$. On the other hand, it is possible to use the dispersion relation and the phonon polarization vectors to construct another frequency function which completely describes the neutron spectra.

Molecular vibration frequencies can also be observed with optical techniques. The frequencies that appear in the infrared or Raman fundamental spectra are those corresponding to motions involving the fluctuations of the dipole moment or the polarizability. In crystals this means that only the transverse optical modes couple with the radiation. In order to obtain information about the entire frequency distribution, it becomes necessary to consider the overtones and combination bands, where multiphonon or anharmonic effects render the selection rules largely inoperative. Alternatively, impurity-induced spectra also provide similar information. For inelastic neutron scattering, there exist no analogous selection rules. Since all frequencies are active, the resulting spectrum usually appears as a continuous band. The conventional method of interpreting optical fundamental spectra is based on the assignment of energy levels. This approach is applicable if the data are line spectra and if one is only concerned with line positions. When many transitions are present a more involved method must be used. This is the situation with most inelastic-neutron-scattering measurements and also with the side bands in the optical data. In such cases, even if the broad peaks can be identified, quantitative interpretation requires taking the band shape explicitly into account. Since intensity analysis invariably involves extensive calculations, the interpretation of neutron spectra is difficult. Equally difficult are attempts to correlate neutron data with infrared results. A direct comparison of corresponding peaks in the two kinds of spectra is certainly useful for purposes of level assignment, but the method is clearly limited and should be used with caution.

The procedure of treating molecular systems as polycrystals with more than one atom per unit cell invariably involves a formidable computation problem. It is only recently that lattice dynamics calculations of this type have been initiated. In the absence of detailed calculations, one has to interpret the data in terms of less systematic, phenomenological treatments. In these studies too the concept of relating neutron spectra to frequency distribution proves to be useful. Since intermolecular interactions are generally much weaker than molecular binding energies, it is reasonable to discuss the various motions in terms of external (translation and rotation) and internal (intramolecular) modes. In experiments using thermal or cold incident neutrons, only the external modes are of interest since excitations of the internal states are energetically not possible. To make the calculation tractable, it has been necessary to assume that interactions between the various degrees of freedom can be ignored; the over-all scattering is then a convolution of individual effects due to translation, rotation, and vibration. The most serious defect of this approach lies in the neglect of translation-rotation interactions. While these effects are obviously appreciable in systems such as hydrogen-bonded substances, the extent to which they can alter interpretation has not been studied.

All existing theoretical descriptions of incoherent neutron scattering by molecular samples have made use of the above decomposition assumption. The problem becomes one of treating separately the translational and rotational motions. By combining the two parts in an appropriate way one obtains an idealized cross section. To compare the result with actual experimental data it is then necessary to fold in the spectrum of incident-neutron energies and the resolution function associated with the energy-detection apparatus. These "experimental" effects vary from one measurement to another, but generally they cause a broadening of the real structure in the spectrum. Another effect that can be quite important and that has not received much attention is multiple scattering. Recent results show that additional work is needed to clarify the extent of the data distortion in experiments using samples of typical thickness.

The formal theory of inelastic neutron scattering by an assembly of atoms and molecules is by now very well developed. There exist in the recent literature excellent and quite comprehensive treatments of the subject. However, most discussions of basic principles are so involved in theoretical detail that the nonexpert has difficulty in obtaining much physical insight without exerting considerable effort. In Chapter 2 we introduce the theoretical basis underlying current descriptions of inelastic neutron scattering. The discussions are specifically intended to orient the uninitiated reader and to provide the framework for calculations and data analysis considered in subsequent chapters.

In Chapter 3, the two basic methods for treating neutron spectra of molecular solids and liquids are described. The first approach, restricted mainly to solids, involves a detailed analysis of vibrational motions in terms of lattice waves. The method is widely used in the study of thermal properties of simple crystalline solids, but its application to molecular substances has not received much attention until recently. In principle, the normal-mode calculation provides a precise dynamic description of the system; however, with few exceptions it has not been carried out in practice because of computational difficulties. The second approach treats the molecule as a basic dynamical unit and therefore emphasizes motions such as center-of-mass translations and rotations. This method is more phenomenological, but it is applicable to any physical system.

The neutron data presented in this monograph are compared whenever possible with the results of Raman and infrared studies. The complementary use of the three techniques has proved valuable in cases where the calculation of the normal modes (both internal and external) of the crystal can be performed and where a least-square adjustment of the force constants has been attempted. The fundamental infrared frequencies can usually be assigned even on the basis of a crude model.

A knowledge of the dispersion curves of the crystal is, however, necessary for assignment of neutron frequencies.

The neutron spectra of various hydrogenous compounds are presented in Chapters 4 through 11. Polymeric molecules are discussed first because extensive normal-mode calculations for them are available and because the neutron data have provided information about the acoustic branches (below 500 cm^{-1}) which in most cases could not be observed directly by optical techniques because of selection rules. These bands, together with some of the optical branches of normal modes (500–1000 cm^{-1}), have been observed (Chapters 4 and 5) for a number of polymers. Stretch-oriented samples of polyethylene and polyoxymethylene have been used to show that the intensity of the acoustic modes (which involve skeletal vibrations of the molecule and are quite sensitive to intermolecular forces) is strongly dependent upon the orientation of the sample with respect to that of the incident or reflected beam, thereby facilitating identification. The molecule of polyglycine has been studied in two conformations (zigzag transplanar and helical) and an attempt has been made to characterize the transitions in terms of changes in thermodynamic properties (Chapter 5). Systems involving strong hydrogen bonds are considered in Chapter 6, where observations of the relative intensities of low-frequency vibrations (0–300 cm^{-1}) in neutron spectra provide some insight into the symmetry of the hydrogen bond.

Chapter 7 offers a good example of the complementary aspects of the neutron and optical techniques. Sum and difference bands in infrared spectra can only be unambiguously assigned if the lattice modes of hydroxides are determined, and for them neutron spectra offered the first measurements of these modes. Torsional (or hindered rotational) motions (Chapter 8) are particularly well suited to neutron-scattering studies because they are generally not observable (because of optical selection rules) by conventional spectroscopic techniques. Intramolecular vibrations of molecules or groups of atoms can also be considered as phonons but with zero dispersion [that is, $\omega(\mathbf{q}) = $ constant]. It has been found that for hydrogenous groups in a molecule (for example, HN_4, CH_3), excitation or de-excitation of torsional modes provides the dominant contribution to the scattered neutron spectrum.

The vibration frequencies of H_2O molecules in a variety of environments are studied in Chapters 9 and 10. Characteristic torsional frequencies (about 400–600 cm^{-1}) and translational frequencies (100–200

cm^{-1}) are observed whether the H_2O molecules are in a cluster (ice), move cooperatively (water, hydrates), are isolated in a matrix, or are in contact with different ions.

In Chapter 11 the results of neutron-scattering measurements for molecules of particular interest are summarized. Hexamethylenetetramine is studied, for example, because it is the simplest organic crystal (cubic, one molecule/unit cell) and the only one for which a normal-mode calculation has been attempted and dispersion curves obtained. Neutron data provide the only experimental means of checking the validity of these calculations. The hydrides are also well suited to neutron-scattering studies because the lattice modes involve direct motions of hydrogen atoms and appear very sharply in the neutron spectra. If hydrogen motions can be represented approximately as those of a harmonic oscillator, it is possible to determine from neutron data the various energy levels corresponding to excited states of the molecules in a harmonic potential well.

Finally, a brief description of the most commonly used neutron-scattering technique is presented in Chapter 12.

General References

The standard reference in neutron diffraction is Bacon [G. E. Bacon, *Neutron Diffraction* (Oxford University Press, London, 1962), 2nd ed.] It contains a thorough discussion of the phenomenon of neutron scattering, the experimental techniques of diffraction measurements, and the study of atomic, molecular, and magnetic structures of matter. The most comprehensive and authoritative exposition on inelastic neutron scattering is a collection of review articles edited by Egelstaff [*Thermal Neutron Scattering*, edited by P. A. Egelstaff (Academic Press, Inc., New York, 1965)]. This work contains extensive lists of references up to about 1965. Another valuable source of original work and useful reviews is the proceedings of a series of symposia sponsored by the International Atomic Energy Agency [*Inelastic Scattering of Neutrons in Solids and Liquids* (International Atomic Energy Agency, Vienna, 1961; 1963, 2 Vols.; 1965, 2 Vols)]. The most recent symposium on this subject was held in Copenhagen in May 1968, the proceedings of which should appear in 1968–1969.

For a more pedagogical treatment of neutron scattering, the survey of Brockhouse *et al.* [B. N. Brockhouse, H. Hautecler, and H. Stiller, in *Interaction of Radiation with Solids*, edited by Strumane *et al.* (North Holland Publishing Company, Amsterdam, 1964)] and the monograph of Turchin [V. F. Turchin, *Slow Neutrons* (Israel Program for Scientific Translation, Jerusalem, 1965)] are both useful.

The literature dealing specifically with neutron studies of solids is large and still rapidly growing. Perhaps the most informative surveys are to be found in summer school lectures [*Phonons and Phonon Interactions*, edited by T. Bak (W. A. Benjamin, New York, 1964); *Phonons in Perfect Lattices and in Lattices with Point Imperfections*, edited by R. W. H. Stevenson (Plenum Press, New York, 1966)]. For an introduction to liquid state investigations using neutrons and other radiation techniques, the monograph by Egelstaff [P. A. Egelstaff, *An Introduction to the Liquid State* (Academic Press, Inc., New York, 1967)] gives a clear account of current developments.

Part II
Theory

2. Basic Principles

In this chapter we develop some of the basic concepts in the theory of inelastic neutron scattering by an assembly of atoms or molecules. We will begin with a brief derivation of the double differential cross section, with emphasis on the ways in which the structure and dynamics of the scattering system manifest themselves in the neutron spectrum. The discussions are intended to provide a theoretical background for later calculations and interpretations of experiment; for this reason they are presented from a relatively general point of view. On the other hand, we do not intend to include all the important theoretical developments in this field.[1-3]

Unlike electromagnetic-radiation scattering processes, the scattering of neutrons can be a coherent or an incoherent phenomenon, depending upon the presence of spin and isotopic effects. The distinction between coherent and incoherent processes is important in the present work since we will be primarily concerned with the study of hydrogenous substances, where incoherent scattering is dominant. In Section 2.1 the coherent and incoherent scattering laws are discussed in some detail, and the specific case of neutron scattering by protons is considered. In certain idealized cases, exact cross section results can be obtained. These are presented in Section 2.2 to illustrate the various features of the scattered-neutron energy spectrum that one can anticipate in an actual measurement. The rigorous results, although idealized, provide limiting behavior which will aid data interpretation. In Section 2.3 the space-time representation of cross sections is introduced. This approach

13

is very useful for the discussion of cross-section behavior in terms of atomic motions. It also facilitates considerably the introduction of physical models in approximate calculations. The Gaussian approximation, which is particularly suited to incoherent-scattering studies, is discussed in Section 2.4. This assumption enables us to define a generalized frequency distribution function, which, as we will show, is an appropriate quantity for the analysis of complex band-type neutron spectra. We conclude the chapter with a discussion of infrared absorption in Section 2.5. The absorption cross section is expressed in an analogous space-time representation, and in this way we emphasize the basic similarities in the two techniques. These similarities will be exploited later when attempts to correlate neutron and infrared spectra are discussed in Chapter 3.

2.1   Coherent and Incoherent Scattering Laws

The purpose of this section is to present cross-section expressions for coherent and incoherent scattering and to discuss a number of general properties. Although neutron-scattering cross sections can be derived in several different ways,[1-3] the results are all basically equivalent. We give here only a summary of the important steps and approximations that lead to the double differential cross sections. For readers desiring a more detailed treatment, a full derivation is given in Appendix A.

To compute the scattering cross section in which a neutron suffers energy and momentum transfers of $E_f - E_i$ and $\hbar(\mathbf{k}_i - \mathbf{k}_f)$, we can begin with a result from time-dependent perturbation theory.[4] The probability per unit time that a scattering event takes place is

$$W(n_0\mathbf{k}_i \rightarrow n\mathbf{k}_f) = \frac{2\pi}{\hbar}\delta(E_n + E_f - E_{n0} - E_i)|\langle n\mathbf{k}_f|U|n_0\mathbf{k}_i\rangle|^2, \quad (2.1)$$

where n_0 and n denote the scatterer initial and final states, and U is the neutron-nuclear interaction potential inducing the transition. We include in Equation 2.1 the delta function to ensure that energy is conserved. Equation 2.1 describes a first-order process, namely a transition arising from direct interaction between neutron and scatterer. For potential scattering at low energies, it is customary to replace the short-range potentials by point interactions, [5, 6]

$$U(\mathbf{r}, \mathbf{R}_1, ..., \mathbf{R}_N) = \frac{2\pi\hbar^2}{m}\sum_{l=1}^{N} a_l\,\delta(\mathbf{r} - \mathbf{R}_l), \quad (2.2)$$

where m is the neutron mass, $\mathbf{r}$ and $\mathbf{R}$ are neutron and nuclear positions, and the scatterer is taken to be a system of N nuclei each labeled by the subscript l. This is the well-known Fermi pseudopotential, which contains as an experimental parameter the bound-nucleus scattering length a. The latter is a measure of the strength of the interaction; equivalently, it gives the magnitude of the spherically outgoing scattered wave. For a scattering system of identical nuclei having zero spin, a is simply a constant. However, fluctuations in the scattering length can occur if the scatterer is an isotopic mixture or if the nuclei have nonzero spin. In such cases the interactions must be treated as isotope- or spin-dependent.

The neutron wave function far from the interaction region (the scatterer) can be represented as plane waves. This means that the matrix element in Equation 2.1 can be written as

$$\langle n_0 \mathbf{k}_i | U | m \mathbf{k}_f \rangle = (2\pi\hbar^2/m)\langle n | \sum_l a_l \exp(i\boldsymbol{\kappa} \cdot \mathbf{R}_l) | n_0 \rangle, \qquad (2.3)$$

where $\boldsymbol{\kappa} = \mathbf{k}_i - \mathbf{k}_f$ and where it is understood that if the nuclei have nonzero spin, $|n_0\rangle$ and $|n\rangle$ must be expanded to include the neutron and nuclear spin states as well. The transition probability W is next converted into a cross section. Since the initial states cannot be prepared nor the final states observed (we ignore the class of polarized neutron experiments here), the result must be summed over all final states and averaged over initial states. The differential cross section per unit incident flux and nucleus becomes

$$\frac{d^2\sigma}{d\Omega dE} = \left(\frac{E_f}{E_i}\right)^{1/2} \sum_{nn_0} P(n_0)\,\delta(E_n - E_{n_0} + \hbar\omega)$$

$$\times \frac{1}{N}|\langle n | \sum_l a_l \exp(i\boldsymbol{\kappa} \cdot \mathbf{R}_l)|n_0\rangle|^2, \quad (2.4)$$

where $\hbar\omega = E_f - E_i$ and where $P(n_0)$ is the probability that the initial scatterer state is n_0. Then

$$P(n) = \exp(-E_n/k_b T)\left[\sum_{n'} \exp(-E_{n'}/k_b T)\right]^{-1}, \qquad (2.5)$$

with k_b being Boltzmann's constant and T the temperature.

Equation 2.4 is a general result that can also be obtained in the Born-approximation formalism (see Appendix A). It can be further reduced by removing the scattering length from the transition-matrix element. Such a reduction is instructive because it illustrates the origin of the two processes encountered in a neutron experiment, coherent and incoherent scattering.[7] The distinction between coherent and incoherent is essential to the understanding of any inelastic-scattering study. Experimentally, the relative importance of each type of process can be estimated from a knowledge of the scattering lengths of the given scatterer.[8] For most substances an appreciable amount of coherent scattering is usually present, and often it is the dominant contribution. However, hydrogen is a very important exception. It is well known that neutron scattering by proton is predominantly incoherent because of spin effects.[9] Theoretically, the analysis of coherent and incoherent spectra involves different considerations. The incoherent problem, thus far, has proved to be more tractable, and for this reason the molecular substances that have been investigated invariably contain a significant amount of hydrogen.

When the scatterer is a system of identical nuclei having zero spin, the constant scattering length a can be taken out of the transition-matrix element. The cross section is then proportional to a^2, and in this case there is no incoherent contribution. Incoherence arises when the scattering length can fluctuate from nucleus to nucleus.[10] We consider first the case of a random mixture of isotopes all having zero spin. The observed cross section now must be interpreted as an isotopically averaged quantity. In principle both the scattering length and the matrix element of the density operator, $\exp(i\boldsymbol{\kappa} \cdot \mathbf{R}_l)$, are affected by the averaging process. Whereas a_l can vary considerably from one isotope to another, we expect the variation in the matrix element to be small if the mass difference between the isotopes is small. The reason is that the latter variation occurs mainly through the mass dependence in the dynamics. Assuming this to be the case, we will only average the scattering length a_l over the various isotopes in the scatterer. If we let

$$F_l = \langle n|\exp(i\boldsymbol{\kappa} \cdot \mathbf{R}_l)|n_0\rangle \qquad (2.6)$$

we can rewrite Equation 2.4 to read

$$\left|\langle n|\sum_l a_l \exp(i\boldsymbol{\kappa} \cdot \mathbf{R}_l)|n_0\rangle\right|^2 = \sum_l a_l^2|F_l|^2 + \sum_{ll'}{}' a_l a_{l'} F_l F_{l'}^*, \qquad (2.7)$$

where the prime over the summation indicates that terms for which $l = l'$ are to be omitted. Now we consider averaging Equation 2.7 over a random isotopic distribution while ignoring the effect on F_l. The first term becomes simply

$$\sum_l \overline{a_l^2}|F_l|^2 = \overline{a^2} \sum_l |F_l|^2, \qquad (2.8)$$

where, because the distribution is random, $\overline{a_l^2} = \overline{a^2}$. If the scatterer contains only one chemical element, $|F_l|^2$ is independent of l and Equation 2.8 gives $N\overline{a^2}|F|^2$. The second term in Equation 2.7 is, upon averaging,

$$\sum_{ll'}' \overline{a_l a_{l'}}\, F_l F_{l'}^* = \overline{a}^2 \sum_{ll'}' F_l F_{l'}^*$$

$$= \overline{a}^2 \Big[|\sum_l F_l|^2 - \sum_l |F_l|^2 \Big]. \qquad (2.9)$$

Thus we have

$$\Big| \langle n | \sum_l a_l \exp{(i\boldsymbol{\kappa}\cdot\mathbf{R}_l)} | n_0 \rangle \Big|^2 = (\overline{a^2} - \overline{a}^2) \sum_l |F_l|^2 + \overline{a}^2 \Big| \sum_l F_l \Big|^2. \quad (2.10)$$

It is conventional to refer to $(\overline{a^2} - \overline{a}^2)$ and $\overline{a}^2$ as a_{inc}^2 and a_{coh}^2, the squares of incoherent and coherent scattering lengths respectively.[8] The significance of this separation will be discussed shortly.

For nuclei with nonzero spin, the scattering length in Equation 2.2 has to be modified to take into account the spin dependence of the interaction. The effect arises because the interactions in which neutron and nuclear spins are parallel are different from those in which the spins are antiparallel. The simplest potential that one can construct in terms of the neutron and nuclear spin operators $\mathbf{s}$ and $\mathbf{I}$ and that does not violate conservation of intrinsic angular momentum is the scalar product $\mathbf{s}\cdot\mathbf{I}$. The interaction is usually taken to be[10]

$$U(\mathbf{r}, \mathbf{R}_1, ..., \mathbf{R}_N) = \frac{2\pi\hbar^2}{m} \sum_l [a_l + b_l\,(\mathbf{s}\cdot\mathbf{I}_l)]\, \delta(\mathbf{r} - \mathbf{R}_l), \qquad (2.11)$$

where a_l and b_l are to be determined experimentally. In this case the matrix elements in Equation 2.4 are to be taken over appropriate spin states with a_l replaced by $a_l + b_l(\mathbf{s}\cdot\mathbf{I}_l)$. The spin part of the matrix element calculation is quite straightforward, and we quote here only the results. What one can show is that the cross section, Equation 2.4, can be expressed as the sum of an incoherent and a coherent contribution:[3, 7, 10]

$$\frac{d^2\sigma}{d\Omega dE} = \frac{d^2\sigma_{\text{inc}}}{d\Omega dE} + \frac{d^2\sigma_{\text{coh}}}{d\Omega dE}, \qquad (2.12)$$

$$\frac{d^2\sigma_{inc}}{d\Omega dE} = \left(\frac{E_f}{E_i}\right)^{1/2} a_{inc}^2 \sum_{nn_0} P(n_0)\, \delta(E_n - E_{n_0} + \hbar\omega)$$

$$\times \frac{1}{N} \sum_l |\langle n|\exp(i\boldsymbol{\kappa} \cdot \mathbf{R}_l)|n_0\rangle|^2, \qquad (2.13)$$

$$\frac{d^2\sigma_{coh}}{d\Omega dE} = \left(\frac{E_f}{E_i}\right)^{1/2} a_{coh}^2 \sum_{nn_0} P(n_0)\, \delta(E_n - E_{n_0} + \hbar\omega)$$

$$\times \frac{1}{N} |\langle n|\sum_l \exp(i\boldsymbol{\kappa} \cdot \mathbf{R}_l)|n_0\rangle|^2, \qquad (2.14)$$

where

$$a_{inc}^2 = \overline{\frac{I+1}{2I+1}a_+^2} + \overline{\frac{I}{2I+1}a_-^2} - a_{coh}^2, \qquad (2.15)$$

$$a_{coh}^2 = \overline{\left(\frac{I+1}{2I+1}a_+ + \frac{I}{2I+1}a_-\right)}^2. \qquad (2.16)$$

In these expressions, the scatterer states n_0 and n no longer include the spin states of the neutron and nuclei. The scattering lengths a_+ and a_- correspond to transitions in which neutron and nuclear spins are parallel and antiparallel, respectively. In the case of hydrogen, a_+ and a_- are known as the triplet and singlet scattering lengths.

Examination of Equations 2.13 and 2.14 reveals the basic difference between the incoherent and coherent cross sections. In the latter, the matrix elements are first summed and then squared. This implies that waves scattered by different nuclei can interfere, and consequently the cross section cannot be computed by simply adding the individual contributions from each nucleus. By contrast, the incoherent cross section is obtained as a sum of individual contributions. This does not mean, however, that the scattering takes place between a neutron and an isolated nucleus, since the matrix element is still determined by the properties of the entire scatterer. The statement has been made that coherent scattering gives information about the collective motions in the system. This is certainly true in view of Equation 2.14. It has also been stated that incoherent scattering gives information about single-particle motions. This statement can be misleading. One should keep in mind that Equation 2.13 is just as much affected by the motions of all the nuclei in the system through interatomic forces, and that the single-particle aspect of σ_{inc} refers only to scattering and not to dynamics. Of course, because of this difference $d^2\sigma_{inc}/d\Omega dE$ and $d^2\sigma_{coh}/d\Omega dE$ reveal quite different dynamic properties of the scatterer. This distinction will be further elaborated upon in later sections.

The preceding discussions show clearly that incoherent neutron scattering can arise from only two sources, isotopic mixture and spin effects. Notice that in the case of zero nuclear spin ($I = 0$), we have the earlier results, $a_{inc}^2 = \overline{a^2} - \bar{a}^2$ and $a_{coh}^2 = \bar{a}^2$. In the monoisotopic case,

$$a_{inc}^2 = \frac{I(I+1)}{(2I+1)^2}(a_+ - a_-)^2, \tag{2.17}$$

$$a_{coh}^2 = \left(\frac{I+1}{2I+1}a_+ + \frac{I}{2I+1}a_-\right)^2. \tag{2.18}$$

Thus if $I = 0$ and the system is monoisotopic, a_{inc} vanishes identically.

Since this monograph deals mainly with scattering by hydrogenous substances, it is of interest to evaluate Equations 2.17 and 2.18 for hydrogen. The "free-atom"* triplet and singlet scattering lengths are[9]

$$a_+ = 0.538 \times 10^{-12} \text{ cm},$$

$$a_- = -2.37 \times 10^{-12} \text{ cm},$$

so that

$$a_{inc}^2 = 1.586 \times 10^{-24} \text{ cm}^2,$$

$$a_{coh}^2 = 0.0357 \times 10^{-24} \text{ cm}^2.$$

We see that the large incoherent cross section ($\sigma_{inc} = 4\pi a_{inc}^2$) of hydrogen is the result of a_+ and a_- being of opposite sign and having magnitudes such that $\frac{3}{4}a_+$ nearly cancels $\frac{1}{4}a_-$. The total bound-nucleus cross section of hydrogen, $\sigma_b = 4(\sigma_{inc} + \sigma_{coh})$, is therefore 81.5×10^{-24} cm^2, and this turns out to be an order of magnitude greater than the cross section of other elements. For example, σ_b of carbon is 5.5×10^{-24} cm^2, of which about 1 per cent is incoherent;[8] for deuterium σ_b and σ_{coh} are 7.5 and 5.4×10^{-24} cm^2, respectively. This is the reason that one normally considers only the incoherent contribution from hydrogen in analyzing neutron spectra of hydrogenous compounds.

For purposes of later development, it is more convenient to express the differential cross section in the form (see Appendix A)

---

* The ratio of free-atom to bound-atom scattering lengths is $M(M+m)^{-1}$, where M is the mass of the atom.

$$\frac{d^2\sigma}{d\Omega dE} = \left(\frac{E_f}{E_i}\right)^{1/2} [a_{\text{inc}}^2 \, S_s(\kappa, \omega) + a_{\text{coh}}^2 \, S(\kappa, \omega)] \qquad (2.19)$$

where

$$S(\kappa, \omega) = \frac{1}{2\pi N} \int_{-\infty}^{\infty} dt \exp(-i\omega t) \sum_{ll'} \langle \exp[i\kappa \cdot \mathbf{R}_l(t)] \exp[-i\kappa \cdot \mathbf{R}_{l'}] \rangle$$

$$(2.20)$$

$$S_s(\kappa, \omega) = \frac{1}{2\pi N} \int_{-\infty}^{\infty} dt \exp(-i\omega t) \sum_{l} \langle \exp[i\kappa \cdot \mathbf{R}_l(t)] \exp(-i\kappa \cdot \mathbf{R}_l) \rangle$$
$$(2.21)$$

Equations 2.20 and 2.21 are known as coherent- and incoherent-scattering laws. In these expressions the symbol $\langle \, \rangle$ denotes an equilibrium ensemble average:

$$\langle A \rangle \equiv \sum_n P(n) \langle n|A|n \rangle \qquad (2.22)$$

or classically

$$\langle A \rangle = \int a^3 R_1 \dots a^3 R_N \, d^3 p_1 \dots d^3 p_N \, F(\mathbf{R}_1 \dots \mathbf{R}_N, \mathbf{p}_1 \dots \mathbf{p}_N) \, A, \quad (2.23)$$

where F is an appropriate equilibrium distribution function. The average is also sometimes called the thermal average. Notice that the nuclear (or atomic) positions $\mathbf{R}_l(t)$ and $\mathbf{R}_{l'}$ appear as time-displaced operators [when written without an argument, $\mathbf{R}_l$ is understood to imply $\mathbf{R}_l(0)$], and in general they do not commute except at $t = 0$. It is clear that in this form the effects of the scatterer dynamics are entirely contained in the intermediate scattering functions

$$\chi(\kappa, t) = \frac{1}{N} \sum_{ll'} \langle \exp[i\kappa \cdot \mathbf{R}_l(t)] \exp(-i\kappa \cdot \mathbf{R}_{l'}) \rangle, \quad (2.24)$$

$$\chi_s(\kappa, t) = \frac{1}{N} \sum_{l} \langle \exp[i\kappa \cdot \mathbf{R}_l(t)] \exp(-i\kappa \cdot \mathbf{R}_l) \rangle. \quad (2.25)$$

It is also important to notice that the scattering laws only depend on the variables κ and ω. If we interpret these as any wave vector and frequency, then S and S_s are functions specified by the scatterer properties and can be discussed without reference to neutron scattering. This is a useful point of view which we will discuss further in Section 2.3.

It is appropriate to mention here a number of general properties of the scattering laws.[11] We first observe the symmetry property*

$$S(-\kappa, -\omega) = \exp(-\hbar\omega/k_b T) \, S(\kappa, \omega), \qquad (2.26)$$

* This is also true for S_s.

which is often called the "detailed balance" condition because it gives a relation between neutron energy gain and loss processes. To demonstrate Equation 2.26, one has only to consider

$$S(\boldsymbol{\kappa}, \omega) = Z^{-1} \sum_{nn_0} \exp\left(-E_{n_0}/k_b T\right) \delta(E_n - E_{n_0} + \hbar\omega)$$

$$\times \left| \langle n | \sum_l \exp\left(i\boldsymbol{\kappa} \cdot \mathbf{R}_l\right) | n_0 \rangle \right|^2 \quad (2.27)$$

with $Z = \sum_n \exp\left(-E_n/k_b T\right)$. If we replace $\boldsymbol{\kappa}$ and ω by $-\boldsymbol{\kappa}$ and $-\omega$ in Equation 2.27, we can by interchanging summation indices n and n_0 obtain Equation 2.26 after a slight rearrangement.

In actual computations, it is more convenient to deal with a symmetric function $S_0(\boldsymbol{\kappa}, \omega)$, which is defined as

$$S_0(\boldsymbol{\kappa}, \omega) = \exp\left(-\hbar\omega/2k_b T\right) S(\boldsymbol{\kappa}, \omega). \quad (2.28)$$

If we consider only isotropic and uniform systems (systems with translational and rotational invariance), then

$$S_0(-\kappa, \omega) = S_0(\kappa, \omega) \quad (2.29)$$

because S becomes a function of κ^2. It turns out that if we calculate $S(\boldsymbol{\kappa}, \omega)$ using classical mechanics instead of quantum mechanics, it will have the property shown in Equation 2.29. This implies that a cross section derived from classical arguments will violate the detailed balance condition unless a correction is applied. This point will be discussed in Section 2.4 where we will consider classical calculations.

Another important property of $S(\boldsymbol{\kappa}, \omega)$ arises from its relation to $\chi(\boldsymbol{\kappa}, t)$, namely,

$$S(\boldsymbol{\kappa}, \omega) = \frac{1}{2\pi} \int_{-\infty}^{\infty} dt \exp\left(-i\omega t\right) \chi(\boldsymbol{\kappa}, t). \quad (2.30)$$

If we denote the frequency moments of S by $\overline{\omega^n}$,

$$\overline{\omega^n}(\boldsymbol{\kappa}) = \int_{-\infty}^{\infty} d\omega \, \omega^n S(\boldsymbol{\kappa}, \omega), \quad (2.31)$$

we can rewrite this expression as

$$\overline{\omega^n}(\boldsymbol{\kappa}) = (-i)^n \frac{\partial^n}{\partial t^n} \chi(\boldsymbol{\kappa}, t) \Big|_{t=0}, \quad (2.32)$$

which follows from integration by parts. Thus the coefficients in a power-series expansion in time of $\chi(\boldsymbol{\kappa}, t)$ are the frequency moments of

$S(\boldsymbol{\kappa}, \omega)$. The significance of this relation is that the moments of S are seen to be static quantities, which in principle we know how to evaluate.[12] The computation of ω^n from Equation 2.32, of course, becomes very complicated for large n, and at present only the first few moments have been obtained. These quantities are useful because they give information about short-time behavior, provide consistency conditions for data analysis, and can be employed in approximate calculations. The first few classical moments are[13]

$$\overline{\omega^0_s}(\boldsymbol{\kappa}) = 1, \tag{2.33}$$

$$\overline{\omega^0}(\boldsymbol{\kappa}) = S(\boldsymbol{\kappa}), \tag{2.34}$$

$$\overline{\omega^2_s}(\boldsymbol{\kappa}) = \overline{\omega^2}(\boldsymbol{\kappa}) = \kappa^2(k_b T/M), \tag{2.35}$$

$$\overline{\omega^4_s}(\boldsymbol{\kappa}) = \frac{\kappa^4 k_b T}{M^2}\left[3k_b T + \frac{n}{3\kappa^2}\int d^3 R\, g(R)\,\nabla^2 V(R)\right], \tag{2.36}$$

$$\overline{\omega^4}(\boldsymbol{\kappa}) = \frac{\kappa^4 k_b T}{M^2}\left[3k_b T + \frac{n}{\kappa^4}\int d^3 R\, g(R)(1 - \cos \boldsymbol{\kappa}\cdot\mathbf{R})\,(\boldsymbol{\kappa}\cdot\mathbf{V})^2 V(R)\right], \tag{2.37}$$

where n is the density. All moments of odd order vanish because classically S and S_s are even functions in ω. The structure factor $S(\boldsymbol{\kappa})$, which is studied in X-ray and neutron diffraction, is $1 + \gamma(\boldsymbol{\kappa})$, where

$$\gamma(\boldsymbol{\kappa}) = \int d^3 r\, \exp\left(-i\boldsymbol{\kappa}\cdot\mathbf{r}\right)g(\mathbf{r}) \tag{2.38}$$

and where

$$g(\mathbf{r}) = \frac{1}{N}\left\langle \sum_{ll'} \delta(\mathbf{r} + \mathbf{R}_l - \mathbf{R}_{l'}) \right\rangle \tag{2.39}$$

is the equilibrium two-particle distribution function. In Equations 2.36 and 2.37 the two-body interatomic potential is denoted by V. It is interesting to note from the moment results that V first appears in the fourth moment (quantum mechanically it first appears in the third moment). This means that the lower moments contain no information about atomic motions. Results for the sixth moments have been obtained recently.[14, 15] These involve integrals of V (and its various derivatives) and the three-particle equilibrium distribution function. Since the latter is not known in general, an approximation is necessary in order to obtain numerical results. Higher-order moments have not been computed; these should involve higher-order distribution functions which would make calculation even more involved.

2.2   Cross Sections of Idealized Systems

The preceding cross section expressions can be directly evaluated if the eigenstates and associated energy levels of the system are known. Unfortunately, explicit computations of transition-matrix elements and evaluation of the thermal average can be carried out analytically only in a few idealized cases. These results, though rigorous, generally are not directly useful in data analysis. On the other hand, they do illustrate the different limiting behavior of the cross section, and for readers not familiar with neutron scattering they are an aid to understanding the more involved calculations that will be discussed later.

Among the various idealized systems amenable to exact calculation, the ideal monatomic gas, the simple harmonic oscillator, and the rigid rotator are the most basic examples. The ideal gas represents a system with a continuous spectrum of energy levels, whereas the oscillator and rotator represent systems with discrete states. In all three cases analytic expressions for the scattering laws have been obtained. We will not describe the detailed calculation here. (The case of the harmonic oscillator, which is of some general interest, is discussed in Appendix B.) Rather, we will simply quote the results and only discuss their significance.

In the case of atoms of a monatomic gas (mass M and temperature T), the intermediate scattering function is[1]

$$\chi(\kappa, t) = \exp\left[-(\kappa^2/2M)(i\hbar t + k_b T t^2)\right]. \tag{2.40}$$

At normal densities ($\sim 10^{19}$ cm^{-3}), the interference terms are smaller by a factor of about 10^4; the contribution to Equation 2.40 therefore comes only from the "self" ($l = l'$) terms. The scattering laws are

$$S(\kappa, \omega) = (4\pi k_b T E_R)^{-1/2} \exp\left[-(\hbar\omega + E_R)^2/4k_b T E_R\right], \tag{2.41}$$

where $E_R = \hbar^2\kappa^2/2M$ is the recoil energy. The scattered-neutron spectrum is simply the Doppler line shape, a skew Gaussian with peak position at approximately $E_f = E_i - E_R$. A spectrum of this type is of limited interest since it is completely specified by the mass of the scattering nucleus and the temperature of the medium. As we will show in Section 2.4, this is the behavior of any physical system at sufficiently high incident energies or momentum transfers. Physically, it corresponds to linear trajectory motions over very short distances.

In the case of a system of independent oscillators (mass M, temperature T, and characteristic frequency ω_0), the incoherent intermediate scattering function is (Appendix B)

$$\chi_s(\kappa, t) = \exp\left[-\tfrac{1}{2}\kappa^2\gamma(t)\right], \qquad (2.42)$$

where

$$\gamma(t) = (\hbar/M\omega_0)[\coth z_0 (1 - \cos \omega_0 t) + i \sin \omega_0 t] \qquad (2.43)$$

and $z_0 = \hbar\omega_0/2k_b T$. The interference terms are purely static (time-independent) because of the assumption that different oscillators are dynamically uncoupled. Since this is a rather unrealistic assumption we will not consider the interference terms. From Equations 2.42 and 2.43 we find

$$S_s(\kappa, \omega) = \exp(-2W) \sum_{n=-\infty}^{\infty} \exp(-nz_0) I_n\left(\frac{\hbar\kappa^2}{2M\omega_0}\operatorname{csch} z_0\right)$$

$$\times \delta(E_i - E_f + n\hbar\omega_0), \quad (2.44)$$

where

$$2W = \frac{\hbar\kappa^2}{2M\omega_0}\coth z_0 \qquad (2.45)$$

and $I_n(x)$ is the modified Bessel function of the first kind. The scattered neutron spectrum is a series of equally spaced lines, each corresponding to a transition in which a definite number of oscillator levels is excited. The $n = 0$ term gives the elastic scattering component, and positive (negative) n terms represent excitation (de-excitation) of the system. Notice that at finite temperatures a process in which a neutron loses energy (down-scattering) always has higher intensity than the corresponding energy-gain (up-scattering) process. In fact, at large κ the $n > 0$ transitions have higher intensities than the elastic component, giving rise to an envelope having the same shape as Equation 2.41.

The factor $\exp(-2W)$ in Equation 2.44 is the Debye-Waller factor familiar in X-ray diffraction studies. Its presence implies that at equilibrium each oscillator generates a thermal cloud by virtue of its vibratory motions. The extent of this cloud is of course temperature dependent, and it does not vanish at $T = 0$ because of zero-point vibrations. The mean-square vibrational amplitude in this case is given by

$$\langle u^2 \rangle = \frac{\hbar}{2M\omega_0}\coth z_0, \qquad (2.46)$$

which is merely a statement of equipartition.

The variation of the oscillator cross section with momentum transfer, or scattering angle θ at fixed incident energy, is typical of most of the observed spectra. As κ (or θ) increases, the elastic portion of the spectrum decreases in intensity, whereas the inelastic region shows an increase. The attenuation is brought about by the Debye-Waller factor, but for the inelastic components this is often more than offset by the increase in the Bessel function. The reason for such behavior lies in the fact that the Bessel function argument is generally small enough that $I_n(x)$ can be represented by the leading terms in a power-series expression. The dependence on κ can be quite striking, particularly in those experiments using cold neutrons. For example, at $\theta = 20°$ the spectrum can show a rather large elastic component and very little inelastic scattering. But at $\theta = 90°$ the structure in the elastic region can disappear completely while new structure appears prominently in the inelastic region.

The oscillator model provides a convenient basis for the discussion of more realistic treatments of neutron scattering by solids. It can be applied to any vibrational motions whether they involve center-of-mass, intramolecular, or librational degrees of freedom. The simple result of a single oscillator can be easily generalized to the problem of a continuous distribution of vibrational frequencies. A systematic approach is to begin with a model of coupled oscillators such as those used in the description of lattice vibrations in solids and to derive the frequency distribution in terms of basic interaction parameters (force constants). The calculation is by no means trivial, but the approach does lead to a very precise and valuable interpretation of the neutron data. We will discuss this problem in more detail in the next chapter.

In the case of a system of rigid spherical rotators (moment of inertia I and temperature T), the incoherent intermediate scattering function is[16]

$$\chi_s(\kappa, t) = \sum_{JJ'} \frac{2J'+1}{2J+1} P(E_J) \exp\left[it(E_J - E_{J'})\right] \sum_{l=|J-J'|}^{J+J'} j^2{}_l(\kappa b), \quad (2.47)$$

where $E_J = \hbar^2 J(J+1)/2I$, b is the bond distance, and $j_l(x)$ is the spherical Bessel function of order l. The time transform gives

$$S_s(\kappa, \omega) = \sum_{JJ'} \frac{2J'+1}{2J+1} P(E_J) \, \delta(E_i - E_f + E_J - E_{J'}) \sum_{l=|J-J'|}^{J+J'} j_l^2(\kappa b). \quad (2.48)$$

The spectrum is now a series of nonuniformly spaced discrete lines. It is interesting to note that χ_s, unlike the gas and oscillator results, is not a Gaussian function of κ. Equations 2.46 and 2.47, like similar expressions for linear, symmetric, and asymmetric molecules, permit an exact treatment of the rotational effects in neutron scattering by molecular gases,[16] provided that rotation-vibration interactions can be ignored. Since these results all contain a number of summations, the computation problem can be quite severe in an actual application. Furthermore, it is difficult to introduce the effects of angular-dependent forces into the calculation. For this reason we will consider in Chapter 3 a classical approach that is more suitable for the study of molecular reorientation processes in condensed matter.

2.3   The Space-Time Representation

The procedure of calculating the differential cross sections from Equations 2.20 and 2.21 is applicable to systems for which the eigenvalues and eigenstates are known explicitly. As we have shown, for simple crystalline solids and gases such calculations are feasible. However, for liquids and for most molecular solids, where we have less than quantitative descriptions of the system dynamics, it has been necessary to employ approximate and phenomenological methods for evaluating the scattering laws. The physical basis of these methods as well as general interpretation of the experimental data are best discussed in terms of the space-time representation of the cross sections. By using essentially the relation between Schrödinger and Heisenberg representations, we can express the transition-matrix elements as time-dependent correlation functions. This in fact was what we did in obtaining Equations 2.20 and 2.21. We describe now a further transformation which allows us to focus attention on the space-time variation of the scatterer and not on the stationary energy levels. This approach is particularly useful for physical interpretation and for formulating approximate calculations.

According to Equation 2.20, the spectrum of scattered neutrons is determined by the time dependence of the intermediate scattering function $\chi(\kappa, t)$, the relation between frequency and time being a Fourier transform. We introduce in a similar manner the position variable $\mathbf{r}$ as conjugate to κ. We formally define a space-time function $G(\mathbf{r}, t)$,[10, 17]

$$S(\kappa, \omega) = \frac{1}{2\pi} \int d^3r \int_{-\infty}^{\infty} dt \, \exp\left[+i(\kappa \cdot \mathbf{r} - \omega t)\right] G(\mathbf{r}, t) \qquad (2.49)$$

or

$$G(\mathbf{r}, t) = \frac{1}{(2\pi)^3} \int d^3\kappa \exp\left(-i\boldsymbol{\kappa}\cdot\mathbf{r}\right)\chi(\boldsymbol{\kappa}, t). \qquad (2.50)$$

Comparison with Equation 2.20 shows that $G(\mathbf{r}, t)$ can also be defined as

$$G(\mathbf{r}, t) = \frac{1}{N} \sum_{ll'} \left\langle \int d^3r'\, \delta(\mathbf{r} - \mathbf{r}' - \mathbf{R}_l(t))\, \delta(\mathbf{r}' - \mathbf{R}_{l'}) \right\rangle, \qquad (2.51)$$

where in quantum-mechanical calculations the noncommutivity of $\mathbf{R}_l(t)$ and $\mathbf{R}_{l'}$ should be noted. If we consider only isotropic and uniform systems,

$$G(\mathbf{r}, t) = \frac{1}{n} \sum_{ll'} \left\langle \delta(\mathbf{r} - \mathbf{R}_l(t))\, \delta(\mathbf{R}_{l'}) \right\rangle, \qquad (2.52)$$

where n is the density. Equation 2.52 shows that $G(\mathbf{r}, t)$ is a density-density correlation function since it is the average of a product of two density operators (at different space and time points) of the type

$$\rho(\mathbf{r}, t) = \sum_{l=1}^{N} \delta(\mathbf{r} - \mathbf{R}_l(t)). \qquad (2.53)$$

Notice also that at time $t = 0$,

$$G(\mathbf{r}, 0) = \delta(\mathbf{r}) + g(\mathbf{r}), \qquad (2.54)$$

where $g(\mathbf{r})$ is given by Equation 2.39. Equations 2.52 and 2.54 suggest that $G(\mathbf{r}, t)$ expresses the density of particles to be found at $\mathbf{r}$ at time t, given that a particle was at the origin at $t = 0$. This interpretation is correct but strictly speaking should be applied only to a purely real quantity, whereas $G(\mathbf{r}, t)$ as defined above is complex. The imaginary part of $G(\mathbf{r}, t)$ has as its origin the quantum-mechanical definition of $G(\mathbf{r}, t)$.[18] Since the scattering law is manifestly real, one finds upon taking the complex conjugate of Equation 2.49 that

$$G^*(\mathbf{r}, t) = G(-\mathbf{r}, -t), \qquad (2.55)$$

a property known as Hermitian symmetry. The fact that $G(\mathbf{r}, t)$ is, in general, a complex quantity makes it difficult to ascribe physical meaning to its behavior. On the other hand, the physical content of this function in the classical limit is quite simple and greatly facilitates phenomenological calculations of $S(\boldsymbol{\kappa}, \omega)$.[19]

A careful discussion of the classical limit of $G(\mathbf{r}, t)$ involves details that are beyond the scope of this monograph.[20] It suffices to say that in the limit as $\hbar \to 0$, we can treat $\mathbf{R}_l(t)$ and $\mathbf{R}_{l'}$ in Equation 2.51 as commuting variables and thus obtain

$$G^c(\mathbf{r}, t) = \frac{1}{N} \sum_{ll'} \langle \delta(\mathbf{r} + \mathbf{R}_l(t) - \mathbf{R}_{l'}) \rangle. \tag{2.56}$$

Since for a given l and l' the position vectors must satisfy a triangular condition, it is clear that we can interpret $G^c(\mathbf{r}, t)$, which is real and even in t, as the density of particles at $\mathbf{r}$ at time t given a particle at the origin at $t = 0$. Observe that for $t = 0$ the particle at the origin can also later contribute to the density at $\mathbf{r}$.

In order to see how G and its classical limit, G^c, are related, we first rewrite Equation 2.52 as

$$G(\mathbf{r} - \mathbf{r}', t) = \frac{1}{n} \langle \rho(\mathbf{r}, t) \rho(\mathbf{r}', 0) \rangle \tag{2.57}$$

or

$$G(\mathbf{r} - \mathbf{r}', t) = \frac{1}{2n} \{\langle [\rho(\mathbf{r}, t), \rho(\mathbf{r}', 0)] \rangle + \langle [\rho(\mathbf{r}, t), \rho(\mathbf{r}', 0)]_+ \rangle\}, \tag{2.58}$$

where $[A, B]$ and $[A, B]_+$ denote commutator and anticommutator, respectively. Equation 2.58 can be expressed as

$$G = G_R + i\hbar G_I \tag{2.59}$$

where

$$G_R(\mathbf{r} - \mathbf{r}', t) = \frac{1}{2n} \langle [\rho(\mathbf{r}, t), \rho(\mathbf{r}', 0)] \rangle, \tag{2.60}$$

$$G_I(\mathbf{r} - \mathbf{r}', t) = \frac{1}{2i\hbar n} \langle [\rho(\mathbf{r}, t), \rho(\mathbf{r}', 0)]_+ \rangle. \tag{2.61}$$

In this form we see that G_R and G_I are both real and have nonzero classical limits, namely, $G_R \to G^c$ and $G_I \to (2n)^{-1} \langle [\rho(\mathbf{r}, t) \rho(\mathbf{r}', 0)]_{\text{PB}} \rangle$, where PB denotes the Poisson bracket. The decomposition shows that in the $\hbar = 0$ limit G is real and equal to the limit of G_R,

$$\lim_{\hbar \to 0} G = \lim_{\hbar \to 0} G_R = G^c. \tag{2.62}$$

We have expressed the real part of $G(\mathbf{r}, t)$ as the average of an anti-commutator that is known to describe fluctuation properties of the system. The imaginary part is given by a commutator that describes the dissipation or response properties. There exists a general relation between the fluctuations and the dissipations, known as the "fluctuation-dissipation" theorem.[21] When applied to our problem, the theorem provides the following relation between G_R and G_I, or their Fourier components:[22]

$$\int_{-\infty}^{\infty} dt \exp(-i\omega t) G_R(\mathbf{r}, t) = i\hbar \coth\left(\frac{\hbar\omega}{2k_b T}\right) \int_{-\infty}^{\infty} dt \exp(-i\omega t) G_I(\mathbf{r}, t). \tag{2.63}$$

Expressions like Equation 2.63 are useful in nonequilibrium statistical mechanics because they allow us to study transport processes in matter and the radiation scattering properties of these systems from a unified point of view. In classical terms, we see that neutron scattering is governed by density fluctuations occurring in the system that is at equilibrium. Transport phenomena such as diffusion, shear flow, and thermal conduction depend on how the system behaves when subjected to a disturbance. According to the classical limit of Equation 2.63,

$$G_I(\mathbf{r}, t) = -\frac{1}{2k_b T} \frac{\partial}{\partial t} G_R(\mathbf{r}, t), \tag{2.64}$$

transport properties should also manifest themselves in the spectrum of inelastically scattered neutrons. There is therefore a great deal of current interest in the use of low-energy neutrons to probe the high-frequency ($\omega \sim 10^{13}$ sec^{-1}) and short-wavelength ($\kappa^{-1} \sim 10^{-8}$ cm) processes in liquids and solids.[3, 23]

Thus far we have discussed only the coherent-scattering law. Analogous results exist for incoherent scattering. Thus

$$G_s^c(\mathbf{r}, t) = \frac{1}{n} \sum_l \langle \delta(\mathbf{r} - \mathbf{R}_l(t) - \mathbf{R}_l) \rangle. \tag{2.65}$$

The classical limit of $G_s(\mathbf{r}, t)$ can be interpreted as the conditional probability density for finding the particle at $\mathbf{r}$ at time t, given that it was at the origin at $t = 0$. Obviously, one has

$$G_s(\mathbf{r}, 0) = \delta(\mathbf{r}). \tag{2.66}$$

It is frequently convenient to decompose $G(\mathbf{r}, t)$ in terms of the "self" correlation function $G_s(\mathbf{r}, t)$ and another function $G_d(\mathbf{r}, t)$:[19]

$$G(\mathbf{r}, t) = G_s(\mathbf{r}, t) + G_d(\mathbf{r}, t). \tag{2.67}$$

The quantity G_d is called the "distinct" correlation function because it specifies a particle at $(\mathbf{r}, t)$ which is different from the one at the origin at $t = 0$. It should be emphasized that whereas the separation of the cross section into coherent and incoherent contributions is fundamentally meaningful because they can be experimentally observed, the separation of G into G_s and G_d is made only for convenience in calculation and interpretation. The procedure has been used mainly in the analysis of the spectra of simple liquids.

In conclusion we mention that the time-dependent density correlation functions G and G_s represent basic dynamic properties of any interacting system of particles. From this point of view, inelastic neutron scattering is only one experimental method with which we can obtain information about these functions. There exist a number of other experimental techniques that can also be used to study density fluctuations.[23] In particular, it should be noted that incoherent inelastic neutron scattering, neutron absorption, and emission and absorption of nuclear γ rays can be discussed in terms of G_s, whereas coherent neutron scattering, Rayleigh scattering of light, and X-ray scattering provide information about G. Even though different experiments effectively measure the same function, the wavelength and frequency ranges involved seldom overlap completely, so that different types of data give complementary results rather than identical information. In Section 2.5 we will emphasize the similarities between neutron scattering and infrared absorption using the space-time representation.

2.4   Gaussian Approximation and the Generalized Frequency Distribution

To illustrate the interpretive and calculational advantages of a classical space-time representation, we consider in this section a number of simple examples.[19] These examples not only are useful for orientation purposes but also serve to motivate an approximation procedure that is of general utility in data analysis. Our attention is directed only at incoherent scattering since the experiments discussed in this monograph are to be interpreted in this context.

We begin with the noninteracting gas for which the exact result has already been given in Section 2.2. In view of our interpretation of $G_s^c(\mathbf{r}, t)$, we have

$$G_s^c(\mathbf{r}, t) = \int d^3 V f(V) \, \delta(\mathbf{r} - \mathbf{V}t), \qquad (2.68)$$

where

$$f(V) = (\pi v_0^2)^{-3/2} \exp\left[-(V/v_0)^2\right] \qquad (2.69)$$

is the Maxwellian velocity distribution and $v_0^2 = k_b T/2M$ is the thermal speed. In writing Equation 2.68 we have assumed the particle at the origin at $t = 0$ to have a velocity distribution given by $f(\mathbf{V})$. This is appropriate because $G_s^c(\mathbf{r}, t)$ is the density correlation function for a system at equilibrium. The integration in Equation 2.68 is trivial; one finds

$$G_s^c(r, t) = (\pi v_0^2 t^2)^{-3/2} \exp\left[-(r/v_0 t)^2\right] \qquad (2.70)$$

and the corresponding scattering law

$$S_s^c(\kappa, \omega) = (\pi \kappa^2 v_0^2)^{-1/2} \exp\left[-(\omega/\kappa v_0)^2\right]. \qquad (2.71)$$

Comparison with Equation 2.41 shows that two factors, $\exp\left(-\hbar\omega/2k_b T\right)$ and $\exp\left(-\hbar^2 \kappa^2/8 M k_b T\right)$, are missing in the classical result. The former is the detailed balance factor discussed in Section 2.1 whereas the latter is the recoil factor to be discussed below.

Calculations for a classical oscillator are also straightforward. In this case one has instead of Equation 2.68 the expression

$$G_s^c(\mathbf{r}, t) = \int d^3 V f(V) \, \delta(\mathbf{r} - \mathbf{R}(t)), \qquad (2.72)$$

where $\mathbf{R}(t)$ is the solution to the equation of motion of a simple oscillator with initial condition $\mathbf{R}(0) = 0$. Thus

$$G_s^c(r, t) = \left[\frac{\omega_0^2}{2\pi v_0^2(1 - \cos \omega_0 t)}\right]^{3/2} \exp\left[\frac{-\omega_0^2 r^2}{2 v_0^2(1 - \cos \omega_0 t)}\right] \qquad (2.73)$$

and

$$S_s^c(\kappa, \omega) = \frac{1}{2} \exp\left[-\tfrac{1}{2}(\kappa v_0/\omega_0)^2\right] \sum_{n=0}^{\infty} (2 - \delta_{on}) I_n\left[\frac{1}{2}\left(\frac{\kappa v_0}{\omega_0}\right)^2\right] \\ \times \left[\delta(\omega + n\omega_0) + \delta(\omega - n\omega_0)\right]. \quad (2.74)$$

Another system for which we can readily determine G_s^c and S_s^c is a system of particles undergoing continuous diffusion. The time correlation function should then satisfy the diffusion equation

$$D\nabla^2 G_s^c(\mathbf{r}, t) = \frac{\partial G_s^c(\mathbf{r}, t)}{\partial t}, \qquad (2.75)$$

where D is the self-diffusion coefficient. The solution corresponding to the initial condition $G_s(\mathbf{r}, 0) = \delta(\mathbf{r})$ is

$$G_s^c(r, t) = (4\pi D t)^{-3/2} \exp\left[-r^2/4Dt\right] \qquad (2.76)$$

which holds for $t \geqslant 0$. Its transform is a Lorentzian function,

$$S_s^c(\kappa, \omega) = \frac{1}{\pi} \frac{D\kappa^2}{\omega^2 + (D\kappa^2)^2}. \qquad (2.77)$$

For a more general description of diffusion one can employ the results developed in the theory of Brownian motion. For example, it is not difficult to give the expression for G_s for particles described by the Langevin equation of motion, and one can show that for large values of t this result reduces to Equation 2.76.[19] However, in this case one no longer has an analytic expression for $S_s^c(\kappa, \omega)$.

The preceding results for a gas, an oscillator, and a diffusion atom have an important property in common. It is observed from Equations 2.69, 2.72, and 2.76 that $G_s^c(r, t)$ in all three cases has a Gaussian dependence on $\mathbf{r}$. This behavior suggests that spatial and time dependence can be separated if we assume the form[19]

$$G_s^c(r, t) = [2\pi W(t)]^{-3/2} \exp[-r^2/2W(t)] \qquad (2.78)$$

or equivalently

$$\chi_s^c(\kappa, t) = \exp[-\kappa^2 W(t)/2]. \qquad (2.79)$$

By writing G_s in this way, we emphasize the fact that the behavior of G_s is completely characterized by a time function $W(t)$, which we will call the "width" function. For the three idealized systems,

$$W(t) = \tfrac{1}{2}(v_0 t)^2 \qquad \text{(gas)}, \qquad (2.80)$$

$$W(t) = (v_0/\omega_0)^2 (1 - \cos \omega_0 t) \quad \text{(oscillator)}, \qquad (2.81)$$

$$W(t) = 2Dt \qquad \text{(diffusing atom)}. \qquad (2.82)$$

The physical meaning of the width function is apparent if one examines the gas or oscillator calculation. One finds that $W(t)$ is simply the mean-square displacement of the particle during a time interval t, that is,

$$W(t) = \tfrac{1}{3} \langle [\mathbf{R}(t) - \mathbf{R}]^2 \rangle. \qquad (2.83)$$

In Figure 2.1 we have sketched the various width functions. It should be noted that the gas and oscillator results are identical for small values of t. This is the correct limiting behavior for any system, since over very short time intervals all motions produce straight-line displacements and hence $W(t)$ varies like t^2. The diffusing atom does not show this behavior because the model assumes continuous diffusion, in which case the particle trajectory is nonlinear at all times. At long times all three models differ. It is not surprising that the gas atom shows the greatest displacement, whereas the oscillator motion is sinusoidal and bounded.

The Gaussian representation of $G_s^c(\mathbf{r}, t)$ is particularly well suited to the discussion of neutron scattering by liquids.[19, 22] Although in general there is no reason to believe that G_s^c always has a Gaussian spatial distribution, the fact that models describing very different types

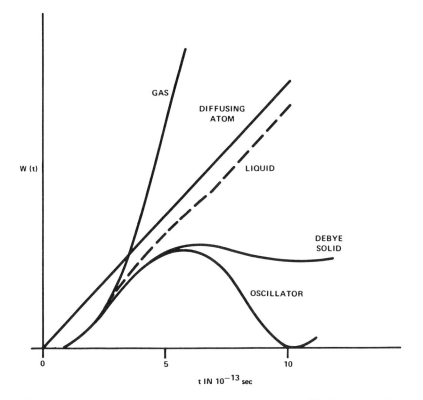

Figure 2.1. Typical mean-square displacement function of a gas, diffusing atom, Debye solid, and oscillator. The dashed line shows the behavior expected in a liquid.

of motion all show this behavior strongly suggests that such an assumption should be a very good approximation even for liquids. Non-Gaussian corrections in liquids[24] and gases[25] have been studied; the effects, present mainly at intermediate times, are generally quite small. In practice, Equation 2.78 is widely used in the analysis of all incoherent spectra. Once the Gaussian assumption is made, the problem reduces to the determination of an appropriate $W(t)$. Since it is not feasible to calculate $W(t)$ for liquids from a molecular theory, a great deal of effort has been directed to the development of model descriptions.[22] We will describe a model in Chapter 3 which has the correct asymptotic behavior at long and short times. In the intermediate-time region ($t \sim 10^{-12}$ sec) the width function is usually thought to behave more like a solid. In Figures 2.1 and 2.2 the liquid curves (dashed lines) satisfy all the general requirements.* It is clear that none of the models discussed thus far are appropriate for liquids in this respect.

The assumption of a Gaussian form for $G_s(\mathbf{r}, t)$ leads to a number of useful quantities and relations for the discussion and calculation of molecular dynamics.[26] Suppose we apply the Gaussian approximation to the quantum-mechanical incoherent-scattering law

$$S_s(\kappa, \omega) = \frac{1}{2\pi} \int_{-\infty}^{\infty} dt \exp(-i\omega t) \exp\left[-\tfrac{1}{2}\kappa^2 \gamma(t)\right], \qquad (2.84)$$

where $\gamma(t)$ is a complex width function yet to be specified. In view of the detailed balance property discussed in Section 2.1, we put

$$S_s(\kappa, \omega) = \exp(-\hbar\omega/2k_bT)S'(\kappa, \omega), \qquad (2.85)$$

where

$$S'(\kappa, \omega) = \frac{1}{2\pi} \int_{-\infty}^{\infty} dt \exp(-i\omega t) \exp\left[-\tfrac{1}{2}\kappa^2 \gamma'(t)\right] \qquad (2.86)$$

with $\gamma'(t) = \gamma(t + i\hbar/2k_bT)$, a real and even function. In Equation 2.86 we integrate by parts twice to obtain

$$S'(\kappa, \omega) = \frac{1}{2\pi} \frac{\kappa^2}{2\omega^2} \int_{-\infty}^{\infty} dt \exp(-i\omega t) \exp\left[-\tfrac{1}{2}\kappa^2 \gamma'(t)\right]$$
$$\times \left\{ \frac{d^2\gamma'}{dt^2} - \frac{\kappa^2}{2}\left(\frac{d\gamma'}{dt}\right)^2 + \dots \right\}. \qquad (2.87)$$

Thus we can define a frequency function $p(\omega)$ as

$$p(\omega) = 2\omega^2 \lim_{\kappa \to 0} \frac{1}{\kappa^2} S'(\kappa, \omega) = \frac{1}{2\pi} \int_{-\infty}^{\infty} dt \exp(-i\omega t) \left(\frac{d^2\gamma'}{dt^2}\right)$$
$$= \frac{1}{\pi} \int_0^{\infty} dt \cos \omega t \left(\frac{d^2\gamma'}{dt^2}\right). \qquad (2.88)$$

This is an interesting result because it provides a means of extrapolating the neutron measurements to determine $p(\omega)$. From Equation 2.88 we have

$$\frac{d^2\gamma'(t)}{dt^2} = 2 \int_0^{\infty} dt \cos \omega t \, p(\omega). \qquad (2.89)$$

Integrating this expression twice and using as boundary conditions $\gamma'(i\hbar/2k_bT) = 0$ and $(d\gamma'/dt)(i\hbar/2k_bT) = i\hbar/M$, we find[26]

$$\gamma'(t) = 2 \int_0^{\infty} dt \frac{p(\omega)}{\omega^2} \left[\cosh(\hbar\omega/2k_bT) - \cos \omega t\right]. \qquad (2.90)$$

---

* See Section 3.2 for actual results for liquid argon.

The relationship between $\gamma'(t)$ and $W(t)$ is given by

$$\gamma'(t) = \gamma'(0) + W(t) \tag{2.91}$$

so that

$$\gamma'(0) = 2 \int_0^\infty d\omega \frac{p(\omega)}{\omega^2} \left[\cosh\left(\hbar\omega/2k_bT\right) - 1\right] \tag{2.92}$$

and

$$W(t) = 2 \int_0^\infty d\omega \frac{p(\omega)}{\omega^2} \left[1 - \cos \omega t\right]. \tag{2.93}$$

That Equation 2.91 expresses the correct relation can be seen from the condition $W(0) = 0$ and from the fact that $d^2\gamma'/dt^2 = d^2W/dt^2$. Thus Equation 2.89 is consistent with Equation 2.93.

It is of interest to observe a relation between the mean-square displacement or width function $W(t)$ and the velocity autocorrelation function $\langle v(t)\,v(0)\rangle$. From Equation 2.83 and using the property of stationary systems one can establish that

$$W(t) = 2 \int_0^t dt'\,(t - t')\ \langle v(t')\,v(0)\rangle \tag{2.94}$$

or

$$\frac{1}{2}\frac{d^2 W(t)}{dt^2} = \ \langle v(t)\,v(0)\rangle. \tag{2.95}$$

Thus the frequency function $p(\omega)$ is simply the Fourier transform (or spectral representation) of the velocity autocorrelation function,

$$p(\omega) = \frac{2}{\pi}\int_0^\infty dt \cos \omega t \ \ \langle v(t)\,v(0)\rangle. \tag{2.96}$$

If we introduce a function $g(\omega)$ such that

$$g(\omega) = \frac{2M}{\hbar\omega}\sinh\left(\frac{\hbar\omega}{2k_bT}\right)p(\omega), \tag{2.97}$$

then

$$\gamma(t) = \frac{\hbar}{M}\int_0^\infty d\omega \frac{g(\omega)}{\omega}\left[\coth\left(\frac{\hbar\omega}{2k_bT}\right)(1 - \cos \omega t) + i \sin \omega t\right]. \tag{2.98}$$

Equation 2.98 shows that in the case of a harmonic solid $g(\omega)$ is precisely the distribution of lattice vibrational frequencies. For a simple oscillator, $f(\omega) = \delta(\omega - \omega_0)$ and we obtain Equation 2.43. It is conventional to call $p(\omega)$ the generalized frequency distribution.* Typical results for the model systems considered previously are shown in

* At high temperatures, $g(\omega) = (M/k_bT)p(\omega)$.

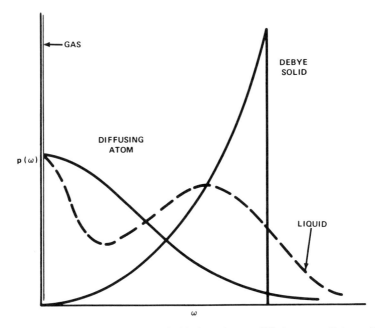

Figure 2.2 Generalized frequency distribution of a gas, diffusing atom, Debye solid. The dashed line shows the behavior expected in a liquid.

Figure 2.2. For a gas $p(\omega)$ is proportional to a delta function $\delta(\omega)$, and for a system of particles obeying the Langevin equation it is a Lorentzian. The zero frequency value of $p(0)$ is of interest,

$$p(0) = 2 D/\pi, \qquad (2.99)$$

where D is the self-diffusion coefficient. The fact that the integral of the velocity autocorrelation function $\langle v(t) v(0) \rangle$ is related to the self-diffusion is another example of the fluctuation-dissipation theorem mentioned in Section 2.3. For the systems shown in Figure 2.2, $p(0)$ is therefore infinite, finite, and zero. Finally, the normalization of $p(\omega)$ is given by

$$\frac{2M}{\hbar} \int_{0}^{\infty} d\omega \frac{p(\omega)}{\omega} \sinh\left(\frac{\hbar\omega}{2k_b T}\right) = 1 \qquad (2.100)$$

which implies that $g(\omega)$ has unit normalization. It should be pointed out that these results are appropriate to a monatomic system. In Chapter 3 we will generalize this approach to a molecular system;

because of the presence of rotational frequencies slight modifications will be required.

In summary we see that in the Gaussian approximation the incoherent-scattering law is

$$S_s(\kappa, \omega) = \exp\left(-\hbar\omega/2k_b T\right)\exp\left[-\tfrac{1}{2}\kappa^2\gamma'(0)\right]S_s^c, \qquad (2.101)$$

$$S_s^c(\kappa, \omega) = \frac{1}{2\pi}\int_{-\infty}^{\infty} dt\, \exp\left(-i\omega t\right)\exp\left[-\tfrac{1}{2}\kappa^2 W(t)\right]. \qquad (2.102)$$

The two exponential factors in Equation 2.101 represent corrections to S_s^c in a cross-section calculation. At high temperatures one can readily show that $\exp\left[-\tfrac{1}{2}\kappa^2\gamma'(0)\right]$ becomes $\exp\left(-\hbar^2\kappa^2/8Mk_bT\right)$, which is the recoil factor known to be missing in a classical treatment of $S_s(\kappa, \omega)$.[20] It is also known that in using S_s^c, quantum corrections of order $\hbar^2$ and higher have been ignored. For most purposes this is not serious since one is unable to compute G_s^c for any physical system in a very precise manner.

2.5   Comparison with Infrared Absorption

From the standpoint of molecular spectroscopy, it is useful to compare the techniques of inelastic neutron scattering and infrared absorption at the level of theoretical considerations and experimental observations. It is generally recognized that in spite of the quite different interactions involved in the two processes, basically the same dynamic properties can be measured with neutron scattering and infrared absorption. Because each method probes a different wavelength-frequency region, it is also to be anticipated that the information derived from the two experiments will more likely be complementary and become identical only in certain special cases. The purpose of this section is to discuss briefly the absorption cross section in a space-time formulation analogous to that used in Section 2.3 and to emphasize in this way the complementary nature of absorption and scattering processes. The connection between neutron scattering and interaction with electromagnetic radiation has been discussed in general terms.[27] Recently, Godron has presented a unified review of the various spectroscopic methods of studying molecular motion using time correlation functions.[28] Initial attempts have been made to correlate quantitatively optical and neutron spectra. Some of the analytical methods will be discussed in Chapter 3. In subsequent chapters dealing with experiments on hydrogenous compounds, comparisons of neutron and infrared absorption results will also be made.

In first-order perturbation theory, the transition probability for the absorption of a photon of frequency ω and polarization ζ is expressed by the equation

$$W(n_0\omega_k\zeta \to n) = \frac{2\pi}{h}\delta(E_{n_0} - E_n + \hbar\omega_k)\,|\langle n0|H''|n_0\omega_k\rangle|^2, \quad (2.103)$$

where the interaction potential is

$$H'' = \frac{e}{mc}\int d^3r\,\mathbf{A}(\mathbf{r})\cdot\sum_{l=1}^{N} Z_l p_l\,\delta(\mathbf{r} - \mathbf{R}_l). \quad (2.104)$$

The particle index runs over both electron and nuclei, Z_l being the charge and $\mathbf{p}_l$ the momentum operator. The vector potential for the radiation field has the usual representation[29]

$$\mathbf{A}(\mathbf{r}) = \sum_{\mathbf{k}'\lambda'}\left(\frac{2\pi\hbar c}{V\omega_{k'}}\right)^{1/2}\epsilon_{\mathbf{k}'\zeta'}\left[b_{\mathbf{k}'\zeta'}\exp\,(i\mathbf{k}'\cdot\mathbf{r}) + b_{\mathbf{k}'\zeta'}^{+}\exp\,(-i\mathbf{k}'\cdot\mathbf{r})\right], \quad (2.105)$$

where $\mathbf{k}'$ is the photon wave vector, $\omega_{k'} = c|\mathbf{k}'|$, $\epsilon_{\mathbf{k}'\zeta'}$ is the corresponding polarization vector, and b and b^{+} are the photon destruction and creation operators, respectively. In order to satisfy the condition $\nabla\cdot\mathbf{A} = 0$, we must require that $\epsilon_{\mathbf{k}'\zeta'}$ be perpendicular to $\mathbf{k}'$; hence Equation 2.105 gives only transverse waves. As in Equation 2.3, the transition-matrix element in Equation 2.103 can be reduced. One finds that

$$\langle n0|H''|n_0\omega_k\rangle = -\frac{2\pi\hbar e}{m\omega_k}\sum_l Z_l\langle n|(\epsilon_{k\zeta}\cdot\mathbf{P}_l)\exp\,(i\mathbf{k}\cdot\mathbf{R}_l)|n_0\rangle, \quad (2.106)$$

and already the similarity with Equation 2.3 is apparent. At this stage it is conventional to invoke the dipole approximation by replacing the momentum (k) representation of the density operator, $\exp[i\mathbf{k}\cdot\mathbf{R}]$, by unity. While this approximation should hold well in infrared absorption, it is certainly not valid in neutron scattering. The transition probability can then be written as

$$W(n_0\omega_k\lambda \to n) = \frac{4\pi^2}{V}\omega_R\,\delta(E_{n_0} - E_n + \hbar\omega_k)\,|\langle n|\epsilon_{k\zeta}\cdot\mathbf{M}|n_0\rangle|^2, \quad (2.107)$$

where $\mathbf{M} = e\sum_l Z_l\mathbf{R}_l$ is the total dipole moment of the scattering medium. The corresponding cross section is

$$\sigma_{IR}(\omega) = \frac{4\pi^2}{hc}\omega\sum_{nn_0} P(n_0)\,\delta(\omega_{n_0} - \omega_n + \omega)\,|\langle n|\epsilon\cdot\mathbf{M}|n\rangle|^2 \quad (2.108)$$

where we have dropped the photon wave vector and polarization index in ω and ϵ. Equation 2.107 can also be expressed as a time integral like Equation 2.20. Assuming that the system is isotropic, we can average over the direction of ϵ and obtain

$$\sigma_{IR}(\omega) = \frac{4\pi^2}{\hbar c}\,\omega\,\frac{1}{6\pi}\int_{-\infty}^{\infty} dt\,\exp(-i\omega t)\sum_{ll'}\,\langle\mathbf{M}_l\cdot\mathbf{M}_{l'}(t)\rangle, \quad (2.109)$$

where $\mathbf{M}_l$ is the dipole moment of the lth molecule. The dipole-dipole correlation function, $\langle\mathbf{M}_l\cdot\mathbf{M}_{l'}(t)\rangle$, also can be expressed in terms of a rotational correlation function appropriate to a particular vibration transition. The use of such time correlation functions in the study of infrared and Raman spectra of liquids and gases has been discussed by Gordon.[30]

The cross-section expressions for coherent neutron scattering (Equation 2.49) and for infrared absorption (Equation 2.109) are seen to be Fourier transforms of certain time correlation functions. In the neutron case it is the Fourier (spatial) component of the density-density correlation function $G(r, t)$, whereas in the radiation absorption case it is the dipole-dipole correlation function. The distinction is that the latter, because of the dipole approximation, is a correlation in displacements, whereas the former is a correlation in the distribution of displacements. If the dipole approximation is not used, one finds instead of Equation 2.109 the expression[27]

$$\sigma_{IR}(k, \omega) = \frac{4\pi^2}{\hbar c}\,\omega\,\frac{1}{6\pi}\int d^3r\,dt\,\alpha(\mathbf{r}, t)\exp[i(\mathbf{k}\cdot\mathbf{r}-\omega t)], \quad (2.110)$$

$$\alpha(\mathbf{r}, t) = n^{-1}\langle\mathbf{J}(\mathbf{r}, t)\cdot\mathbf{J}(0, 0)\rangle, \quad (2.111)$$

$$\mathbf{J}(\mathbf{r}, t) = -\frac{e}{m}\sum_{l=1}^{N} Z_l\mathbf{p}_l\,\delta(\mathbf{r} - \mathbf{R}_l(t)); \quad (2.112)$$

$\alpha(\mathbf{r}, t)$ is a velocity correlation function. These results should be compared with Equation 2.49 with

$$G(\mathbf{r}, t) = n^{-1}\langle\rho(\mathbf{r}, t)\,\rho(0, 0)\rangle \quad (2.113)$$

and $\rho(\mathbf{r}, t)$ is defined as in Equation 2.53. Now both cross sections are double Fourier transforms of space-time functions which reflect the fluctuations of the equilibrium system. For processes that affect the behavior of $\alpha(\mathbf{r}, t)$ and $G(\mathbf{r}, t)$ in the same way, the corresponding dynamic properties can be measured with either radiation or neutrons, the only differences being in the magnitudes of the wavelengths and frequencies involved.

REFERENCES

1. A. C. Zemach and R. J. Glauber, *Phys. Rev.* **101**, 118 (1956).
2. L. S. Kothari and K. S. Singwi, *Solid State Physics* **8**, 109 (1959).
3. *Thermal Neutron Scattering,* edited by P. A. Egelstaff (Academic Press Inc., New York, 1965).
4. Any text on quantum mechanics. For example, L. I. Schiff, *Quantum Mechanics* (McGraw-Hill Book Company, Inc., New York, 1955), p. 199.
5. E. Fermi, *Ricerca Scientifica* **7**, 13 (1936); English translation available as USAEC Report NP–2385, United States Atomic Energy Commission, Washington, D.C.
6. G. Breit, *Phys. Rev.* **71**, 215 (1947); B. A. Lippmann and J. Schwinger, *Phys. Rev.* **79**, 469 (1950); G. C. Summerfield, *Ann. Phys. (N.Y.)* **26**, 72 (1964).
7. D. J. Hughes, *Pile Neutron Research* (Addison-Wesley Publishing Company, Inc., Reading, Mass., 1952); *Neutron Optics* (Interscience Publishers, New York, 1954).
8. D. J. Hughes and J. A. Harvey, *Neutron Cross Sections, BNL–325* (1958) and supplements (1960, 1964, 1965), U.S. Government Printing Office, Washington 25, D.C.
9. R. D. Evans, *The Atomic Nucleus* (McGraw-Hill Book Company, Inc., New York, 1955).
10. V. F. Turchin, *Slow Neutrons* (Daniel Davey & Company, Inc., New York, 1965).
11. M. Nelkin, in *Inelastic Scattering of Neutrons in Solids and Liquids* (International Atomic Energy Agency, Vienna, 1961), p. 3.
12. G. Placzek, *Phys. Rev.* **86**, 377 (1952); A. Rahman, K. S. Singwi, and A. Sjolander, *Phys. Rev.* **126**, 986 (1962).
13. P. G. deGennes, *Physica* **25**, 825 (1959).
14. B. R. A. Nijboer and A. Rahman, *Physica* **32**, 415 (1966).
15. D. Forster, P. C. Martin, and S. Yip, *Phys. Rev.* **170**, 155 (1968).
16. A. Rahman, *J. Nucl. Energy, Pt. A: Reactor Science* **13**, 128 (1961). G. W. Griffing, *Phys. Rev.* **124**, 1489 (1961); **127**, 1179 (1962); S. Yip, Thesis, University of Michigan, 1962 (unpublished).
17. L. van Hove, *Phys. Rev.* **95**, 249 (1954).
18. L. van Hove, *Physica* **24**, 404 (1958); Th. W. Ruijgrok, *Physica* **29**, 617 (1963).
19. G. H. Vineyard, *Phys. Rev.* **110**, 999 (1958).
20. M. Rosenbaum and P. F. Zweifel, *Phys. Rev.* **137**, B271 (1965).
21. R. Kubo, in *Lectures in Theoretical Physics* (Interscience Publishers, New York, 1959).
22. A. Sjolander, Chapter 7 in Ref. 3.
23. P. A. Egelstaff, *Introduction to the Liquid State* (Academic Press, Inc., New York, 1967); also in *Rept. Progr. Phys.* **29**, 333 (1966).
24. A. Rahman, *Phys. Rev.* **136**, A405 (1964); see also Ref. 14.
25. R. C. Desai and M. Nelkin, *Nucl. Sci. Eng.* **24**, 142 (1966). R. C. Desai, *J. Chem. Phys.* **44**, 77 (1966).
26. P. A. Egelstaff and P. Schofield, *Nucl. Sci. Eng.* **12**, 260 (1962). P. Schofield, in *Fluctuation, Relaxation and Resonance in Magnetic Systems,* edited by D. Ter Harr (Oliver and Boyd, Edinburgh, 1962).

27. A. Sjolander, in *Phonons and Phonon Interactions* (edited by T. A. Bak) (W. A. Benjamin, Inc., New York, 1964); M. Blume, in *Symposium on Inelastic Scattering of Neutrons by Condensed Systems,* Brookhaven National Laboratory Report, BNL–940 (C–45), 1965.

28. R. G. Gordon, in *Advances in Magnetic Resonance,* edited by J. S. Waugh (Academic Press, Inc., New York, 1968), Vol. III, p. 1.

29. W. Heitler, *The Quantum Theory of Radiation* (Oxford University Press, Inc., New York, 1954).

30. R. G. Gordon, *J. Chem. Phys.* **43**, 1307 (1965).

3. Methods of Spectral Analysis

The spectrum of neutrons inelastically scattered by a molecular system is usually a continuous band with considerable structure. The individual peaks almost always appear with a good deal of broadening and overlap. In analyzing such data one can remove a certain amount of the broadening and overlap by taking into account the experimental resolution and incident spectrum effects. Some corrections for multiple scattering are frequently necessary, but this effect is not easy to estimate. In fact, a sufficiently general and reliable method of treating multiple scattering has not been developed.

After all the experimental effects have been eliminated, the ideal scattered spectrum is generally still a continuous distribution. Unless we can resolve and identify the various broad peaks, the intensity of the entire band has to be considered. In coherent-inelastic-scattering studies of crystalline solids, it is possible to interpret the experiment in terms of the observed peak positions and their variation with momentum transfer. The same approach is often difficult and not always appropriate for incoherent spectra. In these cases the more complete analysis is that based on the use of generalized frequency distributions.

Two methods of analyzing neutron spectra of molecular solids and liquids will be considered in this chapter. The first method is based on the lattice-dynamics treatment of solids. In this approach the role of frequency distribution functions can be discussed systematically. The second method treats the entire molecule as a dynamical unit. Here more phenomenological arguments are used to introduce the distribution of translation and libration frequencies.

The lattice approach is discussed in Section 3.1. The calculation begins with the adiabatic approximation which enables us to treat the eigenvalues in the electronic problem as interaction potentials in the nuclear problem. By assuming harmonic forces and cyclic boundary conditions, one can obtain a normal-mode representation of the nuclear displacements. The problem then reduces to a system of coupled oscillators, and one is in a position to discuss at least formally the dynamics of the system in terms of dispersion relations and distribution of vibration frequencies. Under these conditions the scattering laws can be evaluated rigorously in the same manner as discussed in Section 2.2. We will not consider the details of the calculation since they have been extensively reviewed in the literature.[1, 2]

The theoretical results in the lattice approach which are most useful for data analysis are the one-phonon cross sections for coherent and incoherent scattering.[1, 3] The coherent cross section gives information about dispersion relations, whereas the incoherent cross section leads to a determination of the phonon frequency distribution under certain conditions. Since the experiments which we will be concerned with are incoherent measurements, our discussions will for the most part be directed at the frequency distribution function. It will be shown that for solids with more than one atom per unit cell, the scattered-neutron spectrum is determined by a function differing fundamentally from the phonon frequency distribution. Whereas the latter can be defined in terms of the dispersion relation, the former involves both the dispersion relation and the polarization vectors of normal modes. The implication is that complete normal-mode calculations are necessary for the detailed analysis of incoherent spectra. The usefulness of neutron-scattering experiments therefore depends upon the feasibility of such computations.

The molecular approach, which is specifically developed for incoherent scattering, is discussed in Section 3.2. The calculations are based exclusively on the use of Gaussian approximation (see Section 2.4), and on the assumption that translation-rotation coupling effects can be ignored. The latter is admittedly a potentially serious source of error in our analysis. Unfortunately, the coupling effects in neutron scattering have not been studied, so it is not clear to what extent this assumption is justified. The separation of center-of-mass and rotational (librational) motions enables us to treat the two types of motion independently. The scattered spectrum is therefore a convolution of the contribution due translational excitations and that due to rotational excitations. It will be seen that rotational effects are relatively more important in the determination of $S_s(\kappa, \omega)$.

The treatment of translational motions depends upon whether the system is a solid or liquid. In the case of solids, one can use the Debye approximation when no detailed information about translational frequencies is available. For liquids we will consider an interpolation model which satisfies all the general requirements mentioned in Section 2.4. This model has no adjustable parameters and leads to simple analytic expressions for the width function, the velocity auto-correlation function, and the generalized frequency distribution. Theoretical methods for treating the rotational correlation function, the counterpart of the width function in Section 2.4, are even more limited. Explicit calculations have been performed only for freely rotating systems or for completely hindered systems. Alternatively, one can make use of the information provided by optical line-shape measurements. This is an interesting approach, and some recent developments will be briefly discussed in Section 3.3. Preliminary considerations show that the method leads to a useful means of quantitatively correlating neutron and infrared absorption spectra. Finally, the neutron data itself can be used to derive the effective frequency distribution due to translations and rotations. This function differs from the sum of translation and rotation components because the latter enters with an enhancement factor which in general is greater than unity. In Chapter 9 we will consider an application of this method.

3.1 The Lattice Approach

One of the most fundamental applications of the inelastic neutron scattering technique is the investigation of atomic motions in solids. This is a most fruitful and important field of research, which is growing in scope and activity. In this section we will describe how the oscillator results (the Einstein model) obtained in Section 2.2 can be modified to interpret actual experiments. Explicit calculations of $S(\kappa, \omega)$ and $S_s(\kappa, \omega)$ using realistic crystal models are too involved for our discussions; we will therefore discuss only the basic results and comment on their physical significance. Readers interested in further details should directly consult the literature.[1, 3, 4]

We consider a general dynamical description of a harmonic crystal.[5] The interaction potential of the system is assumed to be a sum of pair potentials, each having a quadratic dependence on the atomic displacements. During the formal development it is not necessary to specify the nature or the range of the interaction. But in practice one is frequently restricted to only nearest-neighbor interactions because otherwise the number of force constants is too large, making the calculation unwieldy and the results ambiguous. Although various lattice models have been proposed, probably the most widely used is that introduced by Born and von Kármán.[5] The results that we discuss subsequently are those derived with this model.

The lattice vibrations of a crystal are conventionally analyzed in terms of normal-mode solutions to the equations of motion for all the atoms in the solid. These solutions describe the atomic displacements in the form of propagating waves, $\exp{(i\mathbf{q} \cdot \mathbf{x}_l - \omega t)}$, where $\mathbf{x}_l$ is the equilibrium position of the lth atom, $\mathbf{q}$ is the propagation wave vector, and ω is the frequency. Mathematically the problem is essentially a space-time Fourier analysis. The structure of the solid and the use of cyclic boundary conditions lead to certain basic properties for $\mathbf{q}$ and ω and a relation between these quantities. Suppose the solid under consideration is a simple Bravais lattice (one atom per unit cell) containing N atoms each of mass M. In this case there will be N allowed discrete values of $\mathbf{q}$ and three values of ω for each $\mathbf{q}$. The determination of $\mathbf{q}$ and the dispersion (or frequency-wave factor) relation $\omega_j(\mathbf{q})$, where j is the branch index, constitutes a normal-mode analysis. The calculation is straightforward in principle, but in practice it is often a tedious task. In what follows we will assume that the necessary normal-mode results have already been obtained prior to the cross-section calculation. Then one can show that the position of the lth atom can be represented as[5]

$$\mathbf{R}_l(t) = \mathbf{x}_l + \left(\frac{\hbar}{2MN}\right)^{1/2} \sum_\lambda (\omega_\lambda)^{-1} \mathbf{A}_\lambda \{a_\lambda \exp\left[i(\mathbf{q} \cdot \mathbf{x}_l - \omega_\lambda t)\right]$$
$$+ a_\lambda^+ \exp\left[-i(\mathbf{q} \cdot \mathbf{x}_l - \omega_\lambda t)\right]\}, \quad (3.1)$$

where a_λ^+ and a_λ are phonon "creation" and "destruction" operators that have the same properties as the operators a^+ and a in Appendix B. Now $\mathbf{q}$ and ω_λ are the phonon wave vector and frequency, with subscript λ denoting $(\mathbf{q}, j)$ for convenience. The vector $\mathbf{A}_\lambda$ is the unit polarization vector. These vectors are part of the results of a complete normal-mode calculation, and, as we will see, they also play an important role in determining the scattered-neutron spectrum. They are in general complex and satisfy the orthonormality conditions

$$\sum_j A_{ja}^*(\mathbf{q})A_{j\beta}(\mathbf{q}) = \delta_{\alpha\beta}, \quad \sum_\alpha A_{ja}^*(\mathbf{q})A_{j'\alpha}(\mathbf{q}) = \delta_{jj'}, \quad (3.2)$$

where α, β label three orthogonal axes. One frequently speaks of transverse or longitudinal phonons in the sense that $\mathbf{q}$ is perpendicular or parallel to $\mathbf{A}_\lambda$. The summation over $\mathbf{q}$ in Equation 3.1 extends over the N discrete values, and summation over j extends over three branches.*

Once the normal-mode transformation is achieved, the evaluation of the thermal average χ (κ, t) for the Born–von Kármán model proceeds in essentially the same way as discussed in Appendix B. The calculation is equally straightforward because we have already found $\langle b^+(t)b \rangle$ and other similar quantities. Although the entire scattering law $S(\kappa, \omega)$ or $S_s(\kappa, \omega)$ can be obtained, we choose to examine first only certain parts of the cross section because these particular terms are most relevant in the study of lattice dynamics,[6] and all other contributions can be treated as corrections.

It is generally recognized that in crystal-dynamics studies the two most basic quantities are the phonon frequency distribution function $g(\omega)$ and the phonon dispersion relation $\omega_j(\mathbf{q})$. By $g(\omega) \, d\omega$ we mean the number of vibration frequencies that occur in the interval $d\omega$ about ω. These two quantities are clearly related; in fact, a knowledge of $\omega_j(\mathbf{q})$ is sufficient to construct $g(\omega)$ though the inverse process is not possible. Our main concern therefore is to examine how the quantities $g(\omega)$, $\omega_j(\mathbf{q})$, and $A_j(\mathbf{q})$ enter into the scattering-law expressions. Once the pertinent relations are established, one can predict cross-section behavior in terms of normal-mode results, or one can use the neutron experiment to measure normal modes. It has been pointed out earlier that neutron inelastic scattering is presently the only effective method for the direct determination of $\omega_j(\mathbf{q})$; the same can be said, to a large extent, for the frequency distribution $g(\omega)$.

The particular parts of $S(\boldsymbol{\kappa}, \omega)$ and $S_s(\boldsymbol{\kappa}, \omega)$ of interest are the "one-phonon" cross sections.[6, 7] These are the inelastic terms that correspond to $n = \pm 1$ in Equation 2.44. Because correlations in the motion of different atoms are now taken into account, there are one-phonon cross sections for both coherent and incoherent scattering. If we label the one-phonon contribution to $d^2\sigma/d\Omega d\omega$ as $d^2\sigma^{(1)}/d\Omega d\omega$, we can show (see the end of this section) that

$$\frac{d^2\sigma^{(1)}}{d\Omega d\omega} = \left(\frac{E_f}{E_i}\right)^{1/2} \left[a_{\text{inc}}^2 S_s^{(1)}(\boldsymbol{\kappa}, \omega) + a_{\text{coh}}^2 S^{(1)}(\boldsymbol{\kappa}, \omega)\right] \tag{3.3}$$

* The number of branches is three times the number of atoms per unit cell. The present three branches, for example, can consist of one longitudinal mode and two transverse modes.

with

$$S_s^{(1)}(\kappa, \omega) = \exp(-2W) \sum_\lambda \frac{\hbar^2}{4MN} |\kappa \cdot \mathbf{A}_\lambda|^2 \bar{n}(\omega) \exp(-\hbar\omega/2k_bT)$$

$$\times [\delta(\omega + \omega_\lambda) + \delta(\omega - \omega_\lambda)], \quad (3.4)$$

$$S^{(1)}(\kappa, \omega) = \exp(-2W) \sum_\lambda \frac{\hbar^2}{4MN} |\kappa \cdot \mathbf{A}_\lambda|^2 \bar{n}(\omega) \exp(-\hbar\omega/2k_bT)$$

$$\times \frac{(2\pi)^3}{v_a} \sum_\tau [\delta(\kappa + \mathbf{q} - 2\pi\tau) \delta(\omega - \omega_\lambda) + \delta(\kappa - \mathbf{q} - 2\pi\tau) \delta(\omega + \omega_\lambda)], \quad (3.5)$$

where v_a is the volume of a lattice cell (or volume per unit atom),

$$2W = \sum_\lambda \frac{\hbar}{2MN\omega_\lambda} |\kappa \cdot \mathbf{A}_\lambda|^2 \coth\left(\frac{\hbar\omega_\lambda}{2k_bT}\right) \quad (3.6)$$

and

$$\bar{n}(\omega) = \frac{1}{\hbar\omega \sinh(\hbar\omega/2k_bT)}. \quad (3.7)$$

Equation 3.4 is seen to be the generalization of the $n = \pm 1$ terms in Equation 2.44 with $I_1(x)$ being replaced by its small-argument expression. The energy-conserving delta function ensures that only one phonon is absorbed or emitted. The phonon energy is determined by ω_λ, and since any phonon can interact with the neutrons, we have a sum over λ (or over $\mathbf{q}$ and j). Equation 3.5 is the one-phonon coherent scattering law. As in Equation 3.4, energy is conserved in coherent processes, but one has in addition a second delta function corresponding to "momentum" conservation. Because of translational symmetry, coherent processes involving the creation or destruction of a phonon (specified by $\mathbf{q}$ and j) cannot occur unless the momentum transfer κ is equal to $2\pi\tau - \mathbf{q}$, where τ is the reciprocal lattice vector.

The existence of the second delta function in $S^{(1)}$ means that coherent processes are fundamentally different from incoherent processes. The condition

$$\mathbf{k}_i - \mathbf{k}_f \pm \mathbf{q} = 2\pi\tau \quad (3.8)$$

is the reason that $\omega_j(\mathbf{q})$ can be measured by coherent scattering and not by incoherent scattering. Experimentally, $\mathbf{k}_i$, $\mathbf{k}_f$, and τ can all be fixed, so this condition permits the measurement of $\mathbf{q}$. At the same time the energy delta function determines the value of $\omega_j(\mathbf{q})$, and thus the dispersion relation can be constructed. The determination of phonon dispersion relations in crystals is to date the most significant and the most extensive contribution of the neutron technique. The availability of dispersion-curve measurements has made possible the adjustment of force constants in normal-mode calculations. In a number of cases, a set of force constants is obtained[8] by adjusting the calculated $\omega_j(\mathbf{q})$ to agree with the observed values.

The one-phonon incoherent scattering is of interest because Equation 3.4 allows us to measure the vibration frequency distribution $g(\omega)$. This can be seen from the following argument. The distribution of $\mathbf{q}$ values in the Brillouin zone is always sufficiently dense that the summation over $\mathbf{q}$ can be replaced by an integral. The scattering law becomes

$$S_s^{(1)}(\boldsymbol{\kappa}, \omega) = \exp(-2W) \frac{\hbar^2}{4M} \bar{n}(\omega) \frac{V}{(2\pi)^3 N} \int d^3q \sum_{j=1}^{3} |\boldsymbol{\kappa} \cdot \mathbf{A}_j(\mathbf{q})|^2$$
$$\times [\delta(\omega - \omega_j(\mathbf{q})) + \delta(\omega + \omega_j(\mathbf{q}))], \quad (3.9)$$

where the integral extends over a unit cell in reciprocal space and where $V = Nv_a$ is the volume of the lattice. Suppose now that the lattice has cubic symmetry; this means that[6]

$$\int d^3q\, F_j(\mathbf{q}) [\boldsymbol{\kappa} \cdot \mathbf{A}_j(\mathbf{q})]^2 = \frac{\kappa^2}{3} \int d^3q\, F_j(\mathbf{q}) \qquad (3.10)$$

for arbitrary $\boldsymbol{\kappa}$ and $F_j(\mathbf{q})$. We can formally define the phonon frequency distribution function as

$$g(\omega) = \frac{v_a}{(2\pi)^3} \frac{1}{3} \sum_j \int d^3q\, \delta[\omega - \omega_j(\mathbf{q})] \qquad (3.11)$$

where $g(\omega) = g(-\omega)$. Then

$$S_s^{(1)}(\boldsymbol{\kappa}, \omega) = \frac{\hbar^2 \kappa^2}{4M} \bar{n}(\omega)\, g(\omega) \exp[-(2W + \hbar\omega/2k_bT)], \qquad (3.12)$$

where

$$\exp(-2W) = \exp\left[-\frac{\hbar\kappa^2}{2M} \int d\omega \frac{g(\omega)}{\omega} \coth\frac{\hbar\omega}{2k_bT}\right] \qquad (3.13)$$

Since the volume of a reciprocal unit cell is $(2\pi)^3/v_a$, Equation 3.11 gives the normalization

$$\int_0^\infty d\omega\, g(\omega) = 1. \qquad (3.14)$$

Another useful relationship that can be derived from Equation 3.11 is

$$g(\omega) = \sum_j g_j(\omega), \tag{3.15}$$

where

$$g_j(\omega) = \frac{v_a}{(2\pi)^3} \frac{1}{3} \int d^3q \, \delta[\omega - \omega_j(\mathbf{q})] \tag{3.16}$$

and, after a transformation,[9]

$$g_j(\omega) = \frac{v_a}{(2\pi)^3} \frac{1}{3} \int_{S(\omega)} dS \, |\nabla_q \omega_j(\mathbf{q})|^{-1}, \tag{3.17}$$

where $S(\omega)$ is a surface of constant frequency $\omega = \omega_j(\mathbf{q})$. Equation 3.17 shows how the frequency distribution can be constructed once the dispersion relation is known.[10] In practice it turns out to be more convenient to use the "root-sampling" method which is well suited for calculations using a large digital computer. In the one-dimensional case (for example, isolated chain of polyethylene), Equation 3.15 reduces to

$$g(\omega) = \frac{v_a}{(2\pi)} \left| \frac{d\omega(q)}{dq} \right|^{-1}_{\omega(q)=\omega} \tag{3.18}$$

It is observed in Equation 3.17 or 3.18 that $g(\omega)$ in general contains a number of singularities where $\nabla\omega_j(\mathbf{q}) = 0$. In the neighborhood of these critical frequencies one can expect a high intensity of scattering since these are the regions with the largest density of phonon states.[10]

Equation 3.12 shows that the one-phonon incoherent cross section is proportional to $g(\omega)$. Consequently, a measurement of the energy distribution of the scattered neutrons yields directly the frequency distribution, provided that an approximate form for $g(\omega)$ is first used to remove the Debye-Waller factor and assuming that higher-order phonon contributions are small or can be eliminated. Because $g(\omega)$ also appears in $\exp(-2W)$, an approximate form for $g(\omega)$ must first be assumed; this is usually taken to be a Debye spectrum. The difficulty is not serious if the momentum transfer is smaller or if the experiment is carried out at a low temperature. When the quantity $2W$ is not small compared to unity, the effects of two and more phonon processes in the spectrum become significant. Although there exist methods for treating multiphonon contributions,[11] they are not always used in actual analysis for reasons of computational simplicity. Multiphonon effects can be experimentally minimized by carrying out the measurements at low temperatures and using low-incident-energy neutrons and small scattering angles. Even when appreciable multiphonon scattering is present in the data, the observation of one-phonon transitions may still be possible. This is because multiphonon processes generally give rise to a relatively smooth inelastic spectrum. In order to extract the frequency distribution in this case, an iterative procedure is necessary.

The most severe limiting factor with respect to the foregoing arguments is that Equation 3.12 is, strictly speaking, not applicable to noncubic lattices or crystals with more than one atom per unit cell. For these cases, which are of more interest in molecular spectroscopy, one cannot eliminate the polarization vectors in Equation 3.9; consequently $S_s^{(1)}$ is not expressible in terms of $g(\omega)$ and has a nontrivial dependence on the direction of scattering. On the other hand, the neutron energy distribution in any given direction still exhibits singular behavior at critical frequencies that are direction independent.[10]

Although Equation 3.12 remains unchanged for cubic crystals in powder form, some simplification is possible when one considers noncubic polycrystalline substances. The average over orientations in Equation 3.9 is equivalent to averaging over all directions of κ. The polycrystalline scattering law result therefore can be written as

$$S_s^{(1)}(\kappa, \omega) = \frac{\hbar^2 \kappa^2}{4M} \bar{n}(\omega)\, G(\omega)\, \exp\left[-(2W + \hbar\omega/2k_b T)\right], \quad (3.19)$$

where

$$G(\omega) = \frac{v_a}{(2\pi)^3} \frac{1}{3} \sum_j \int d^3q\, |\mathbf{A}_j(\mathbf{q})|^2\, \delta[\omega - \omega_j(\mathbf{q})]$$

$$= \frac{v_a}{(2\pi)^3} \frac{1}{3} \sum_j \int_{S(\omega)} dS\, |\mathbf{A}_j(\mathbf{q}')|^2\, |\nabla \omega_j(\mathbf{q}')|^{-1}, \quad (3.20)$$

with $\mathbf{q}$ representing wave vectors on the surface $S(\omega)$. We may regard $G(\omega)$ as an effective frequency distribution, and this is the function that one can measure with incoherent neutron scattering. Even though one cannot display an explicit relation between $g(\omega)$ and $G(\omega)$, comparison of Equations 3.17 and 3.20 shows clearly the role played by the polarization vectors.

Our discussion thus far is limited to monatomic solids. In order to extend these results to neutron experiments on hydrogenous compounds, it is necessary to consider lattices with more than one atom per unit cell. Suppose there are n atoms per unit cell and the atoms are labeled by an index μ. The displacement of the μth atom in the lth unit cell can be written[5] in a form analogous to Equation 3.1,

$$\mathbf{R}_l^\mu(t) = \mathbf{x}_l + \frac{\hbar}{2nM_\mu N} \sum_\lambda \omega_\lambda^{-1/2} \mathbf{A}_\lambda^\mu \{ b_\lambda \exp\left[-i(\mathbf{q} \cdot \mathbf{x}_l - \omega_\lambda t) \right]$$
$$+ b_\lambda^+ \exp\left[-i(\mathbf{q} \cdot \mathbf{x}_l - \omega t) \right] \}, \quad (3.21)$$

where $\mathbf{x}_l$ now locates the cell and both mass M and polarization vector $\mathbf{A}$ depend on the atom index. The one-phonon incoherent cross section becomes

$$\frac{d^2\sigma_{inc}^{(1)}}{d\Omega d\omega} = \left(\frac{E_f}{E_i}\right)^{1/2} \frac{\hbar \bar{n}(\omega)}{2nN} \sum_{\lambda, \mu} (a_{inc}^\mu)^2 M_\mu^{-1} |\boldsymbol{\kappa} \cdot \mathbf{A}_\lambda^\mu|^2 \exp\left(-2W_\mu\right)$$
$$\times \left[\delta(\omega - \omega_\lambda) + \delta(\omega + \omega_\lambda) \right], \quad (3.22)$$

where W is a similar modification of Equation 3.6. The incoherent scattering length a_{inc} also depends on the atom index, and the sum over μ includes all the atoms in a unit cell. The expression for a powder is

$$\frac{d^2\sigma_{inc}^{(1)}}{d\Omega d\omega} = \left(\frac{E_f}{E_i}\right)^{1/2} \frac{\hbar^2 \kappa^2 \bar{n}(\omega)}{4M} \sum_\mu (a_{inc}^\mu)^2 \, G_\mu(\omega) \exp\left[-(2W_\mu + \hbar\omega/2k_bT), \right.$$
$$(3.23)$$

with

$$G_\mu(\omega) = \frac{M}{M_\mu} \frac{v_a}{(2\pi)^3} \frac{1}{3} \sum_{j=1}^{3n} \int_{S(\omega)} dS \left| \mathbf{A}_j^\mu(\mathbf{q}') \right|^2 \left| \boldsymbol{\nabla} \omega_j(\mathbf{q}') \right|^{-1}. \quad (3.24)$$

Notice that j takes on $3n$ values corresponding to the $3n$ branches of the dispersion relation. Equation 3.23 is interesting because it shows that even for a polycrystalline substance with more than one atom per unit cell, the one-phonon incoherent cross section is a sum of effective frequency distributions for individual atoms each weighted essentially according to the square of the scattering length. If we consider only the hydrogen contribution, the cross section is proportional to G_H, the effective frequency distribution for a hydrogen atom.* If the hydrogen in the unit cell are not dynamically equivalent, G_H is different from one hydrogen to the other. It is clear that $G_\mu(\omega)$ is not easily related to $g(\omega)$, the true frequency distribution function. However, in a complete normal-mode analysis one knows $\mathbf{A}_j^\mu(\mathbf{q})$ as well as $\omega_j(\mathbf{q})$; consequently, both $g(\omega)$ and $G(\omega)$ can be computed. Once $G_\mu(\omega)$ is obtained, one is in a position to give a complete analysis of the incoherent spectra. One can also anticipate that $G_\mu(\omega)$ will exhibit singular behavior at the critical frequencies of $g(\omega)$; however, the various intensities can be

---

* In subsequent discussions we denote $G_H(\omega)$ simply as $G(\omega)$ since the systems considered will all have more than one atom per unit cell.

significantly different in the two distributions. The effects of the polarization vectors have only recently been investigated; later we shall discuss the case of polyethylene, for which such calculations have been carried out.[12]

Attempts have been made to measure $g(\omega)$ experimentally for simple solids.[8] The element vanadium, an almost purely incoherent scatterer, has been studied several times. As a check of the derived frequency spectrum, other experimental quantities can be computed. The most familiar application in this connection is the prediction of thermo-dynamic quantities and their temperature variation. Since these are integrated quantities, the dependence on $g(\omega)$ is not expected to be very sensitive. In the next section we shall indicate a method for deriving $g(\omega)$ for molecular solids from a neutron measurement. Even though the theoretical basis for this approach is phenomenological, com-parison of the calculated thermodynamic properties with observations shows surprising agreement.

In concluding this section we show how the one-phonon result follows from the formalism of Section 2.1. We confine our attention to a cubic simple Bravais lattice with one atom per unit cell. The inco-herent-scattering law then takes the form

$$S_s(\kappa, \omega) = \frac{1}{2\pi} \int_{-\infty}^{\infty} dt \exp(-i\omega t) \exp\left[-\tfrac{1}{2}\kappa^2\gamma(t)\right], \qquad (3.25)$$

where $\gamma(t)$ is given by Equation 2.98. If we expand the quantity $\exp\left[-\tfrac{1}{2}\kappa^2\gamma(t) + 2W\right]$, with $2W$ defined as in Equation 3.13, in a power series in κ^2, then to order κ^2,

$$S_s(\kappa, \omega) = \exp(-2W)\left[\delta(\omega) + \frac{\hbar\kappa^2}{4M\omega} \frac{\exp(-\hbar\omega/2k_bT)}{\sinh(\hbar\omega/2k_bT)} g(\omega)\right]. \qquad (3.26)$$

The first term is the elastic-scattering component and the second term is seen to be the one-phonon result shown in Equation 3.12. The higher-order terms not shown in Equation 3.26 constitute the multi-phonon contributions. Equation 3.25 can be used directly to obtain the cross section in terms of $g(\omega)$; if the multiphonon terms themselves are of direct interest, convenient expressions for explicit calculations have been derived by Sjolander.[11] Notice that in terms of the general-ized frequency distribution introduced in Section 2.4 (see Equation 2.97), the one-phonon cross section is

$$S_s^{(1)}(\kappa, \omega) = \frac{\kappa^2}{2\omega^2} p(\omega) \exp\left[-(2W + \hbar\omega/2k_bT)\right]. \qquad (3.27)$$

Finally we discuss briefly the anharmonic effects in neutron scattering.[13] Theoretical investigations have shown that anharmonic forces introduce non-Gaussian corrections to $\chi(\kappa, t)$ as well as modifications in the interpretation of the width function expression, Equation 2.98. The former are relatively small, however, so to a good approximation a Gaussian form of χ with a different width function can still be used. The inclusion of cubic and higher-order terms in the interatomic potential enables phonons of different states to interact; in a perturbation-theoretic sense, the state of a phonon then becomes a mixture of the states in the harmonic approximation. A phonon thus loses its well-defined energy, which means that it has a finite lifetime, and that the peaks in the spectrum are broadened. Moreover, anharmonic effects cause a shift, usually to lower frequency, of the peak position. Because of computation difficulties associated with any quantitative estimates, anharmonic effects are usually assumed to be small and are ignored in the interpretation of neutron spectra of solids. This assumption is justified provided that the measurements are carried out at temperatures low compared to the melting point and at small momentum transfers.

3.2 The Molecular Approach

In molecular systems the intermolecular interactions are usually much weaker than the intramolecular forces. It is therefore reasonable to classify the motions in terms of external modes (center-of-mass translations and rotations) and internal modes. For thermal- and cold-neutron-scattering experiments, only the external models are of interest since excitations of the internal states are energetically unlikely for most systems. To facilitate the calculation we assume that translation-rotation couplings can be ignored. This is admittedly a drastic approximation (which is made in all existing calculations), and it constitutes a source of error of essentially unknown magnitude. We will be concerned only with incoherent scattering from hydrogenous substances. Since the hydrogen scattering cross section is large and almost entirely incoherent (see Section 2.1), we will treat only the hydrogen contribution. The cross section per hydrogen is therefore*

---

* Notice that we have suppressed the subscript s for simplicity. All the scattering quantities referred to in this section pertain to incoherent scattering.

$$\frac{d^2\sigma_{\text{inc}}}{d\Omega d\omega} = \left(\frac{E_f}{E_i}\right)^{1/2} (a_{\text{inc}}^{\text{H}})^2 \frac{1}{2\pi} \int_{-\infty}^{\infty} dt \exp\left(-i\omega t\right)\chi(\kappa, t) \qquad (3.28)$$

with $a_{\text{inc}}^{\text{H}}$ as given in Section 2.1. If other atoms in the molecule also have appreciable incoherent cross sections, one has additional terms in the cross section expression similar to Equation 3.28 and multiplied by the appropriate scattering length. The incoherent intermediate scattering function is

$$\chi(\kappa, t) = \langle \exp\left[i\kappa \cdot \mathbf{r}(t)\right] \exp\left(-i\kappa \cdot \mathbf{r}\right)\rangle \qquad (3.29)$$

where $\mathbf{r}$ is the hydrogen position in a laboratory coordinate system.

To separate molecular translations and rotations we write

$$\mathbf{r} = \mathbf{R} + \mathbf{b} + \mathbf{u}, \qquad (3.30)$$

where $\mathbf{R}$ is the molecular center-of-mass position, $\mathbf{b}$ the hydrogen equilibrium position, and $\mathbf{u}$ the instantaneous displacement. Ignoring couplings between these position vectors, we see that $\mathbf{R}$, $\mathbf{b}$, and $\mathbf{u}$ are associated, respectively, with the translational, rotational, and intramolecular vibrational motions of a molecule. The assumption that the various components of $\mathbf{r}$ are dynamically independent implies that χ is approximated by a product of three scattering functions,

$$\chi \simeq \chi_T \chi_R \chi_V \qquad (3.31)$$

where

$$\chi_T(\kappa, t) = \langle \exp\left[i\kappa \cdot \mathbf{R}(t)\right] \exp\left(-i\kappa \cdot \mathbf{R}\right)\rangle, \qquad (3.32)$$

$$\chi_R(\kappa, t) = \langle \exp\left[i\kappa \cdot \mathbf{b}(t)\right] \exp\left(-i\kappa \cdot \mathbf{b}\right)\rangle, \qquad (3.33)$$

$$\chi_V(\kappa, t) = \langle \exp\left[i\kappa \cdot \mathbf{u}(t)\right] \exp\left(-i\kappa \cdot \mathbf{u}\right)\rangle. \qquad (3.34)$$

Since the excitation of intramolecular vibrations is negligible in low-energy scattering experiments, it is sufficient to replace χ_v by its zero-point vibration expression,

$$\chi_v(\kappa, t) \simeq \exp\left(-\kappa^2 \langle u^2 \rangle\right), \qquad (3.35)$$

where $\langle u^2 \rangle$ is the mean-square vibrational amplitude. This is a static quantity; therefore it has only a slight effect on the scattered-neutron spectrum at low energies.

The treatment of χ_T depends on whether the molecular system is in a solid or liquid phase. For a solid we can employ the results previously discussed, namely,

$$\chi_T(\kappa, t) = \exp\left[-\tfrac{1}{2}\kappa^2 W_T(t)\right], \tag{3.36}$$

$$W_T(t) = \gamma_T(t) - \gamma_T(0), \tag{3.37}$$

$$\gamma_T(t) = \int_0^\infty d\omega \, g_T(\omega)\Lambda(\omega, t), \tag{3.38}$$

$$\Lambda(\omega, t) = \frac{\hbar}{\omega}\left[\coth\left(\hbar\omega/2k_bT\right)\cos\omega t - i\sin\omega t\right]. \tag{3.39}$$

The translation frequency distribution g_T is specified separately. If this is not known from normal-mode calculations, a simple approximation can be made using a Debye spectrum

$$g_T(\omega) = 3\omega^2/\omega_D^3, \tag{3.40}$$

where ω_D is the Debye frequency. Alternatively, two Debye spectra may be used to account for different longitudinal and transverse modes. For liquids, χ_T is still given by Equation 3.36 in the Gaussian approximation (see Section 2.4). The problem is to find an appropriate width function $W_T(t)$ having a minimum number of parameters.

The calculation of $W_T(t)$ for liquids has received considerable attention because inelastic neutron scattering provides a rather unique means of observing directly the atomic motions in liquids.[14] Given a measurement of the incoherent cross section one can in principle extract from the data the mean-square displacement function, $W_T(t)$, or the velocity autocorrelation function, $\langle v(t)v(0)\rangle$.[15] This is a difficult process in practice and so far only qualitative results have been obtained.[16] From these studies and also from molecular-dynamics computer experiments,[17] it is known that at short times ($t \lesssim 5 \times 10^{-13}$ sec) the motion of a liquid atom is oscillatory like that of a solid, while at long times ($t \gtrsim 2 \times 10^{-12}$ sec) the motion becomes diffusive. These asymptotic behaviors have been briefly discussed in Section 2.4. Any vibratory component in the motions can be expected to give rise to a negative $\langle v(t)v(0)\rangle$ over some time interval, or, equivalently, will appear in $p(\omega)$ as a peak away from $\omega = 0$. The diffusive component, always present in a liquid, would lead to a linear growth of $W_T(t)$ with time. The finite diffusion coefficient is represented by the nonzero value of $p(0)$. These characteristics clearly manifest themselves in the computer results, and to a lesser degree they also have been observed in the neutron data. We now describe briefly the results of a model that exhibits these characteristics and has no adjustable parameters. This model has been used recently to study liquid argon and other simple liquids with some success.

Because the model we are describing satisfies the general requirements at long and short times, we will call it the interpolation model.[18] Physically it corresponds to treating the atom as moving in a potential (or cage) which is itself relaxing in time. Thus initially the atom undergoes oscillatory motions, but at long times the potential gives rise to a frictional force so that the atom is able to diffuse. The potential is characterized by an oscillation frequency ω_0 which, however, is expressible in terms of the mean-square force on the atom, that is,

$$\omega_0^2 = \int_{-\infty}^{\infty} d\omega \, \omega^2 f(\omega)$$

$$= \frac{1}{3M} \langle \nabla^2 V \rangle, \tag{3.41}$$

where V is the two-body potential. This expression can be evaluated once V and the two-particle equilibrium distribution function $g(r)$ are known;[19] therefore, ω_0 should not be considered an adjustable parameter. The diffusive motion of the atom is characterized by the relaxation time τ_0 of the potential. This quantity can be expressed in terms of ω_0 and the observed self-diffusion coefficient D as

$$\tau_0 = \frac{M}{k_b T} \frac{1}{\omega_0^2 D}. \tag{3.42}$$

Like ω_0, τ_0 is not an adjustable parameter. The advantage of the interpolation model, aside from its having no free parameters, is that it leads to analytical expressions for $\omega_T(t)$, $\langle v(t)v(0)\rangle$, and $p(\omega)$. As explained in Section 2.4 these quantities are all interrelated. Thus

$$\langle v(t)v(0)\rangle = \exp{(t/2\tau_0)} \left[\cos \Omega t + (1/2\tau_0\Omega) \sin \Omega t\right], \tag{3.43}$$

$$p(\omega) = \frac{2}{\pi} \frac{\omega_0^2/\tau_0}{(\omega^2 - \omega_0^2)^2 + (\omega/\tau_0)^2}, \tag{3.44}$$

$$\frac{M}{2k_b T} W_T(t) = \frac{\omega_0^2 - 1/\tau_0^2}{\omega_0^4} + \frac{1}{\omega_0^2\tau_0}t$$

$$+ \omega_0^{-4} \exp{(-t/2\tau_0)} \left[\left(\frac{1}{8\tau_0^3\Omega} - \frac{3\Omega}{2\tau_0}\right) \sin \Omega t - \left(\omega_0^2 - \frac{1}{\tau_0^2}\right) \cos \Omega t\right], \tag{3.45}$$

where $\Omega^2 = \omega_0^2 - 1/4\tau_0^2$. We can compare Equations 3.43–3.45 with the idealized models discussed in Section 2.4. These functions have been computed for liquid argon and are shown in Figures 3.1, 3.2, and 3.3.[18] For comparison we show also the "exact" results as obtained from computer experiments;[17] the agreement is at least qualitative. When

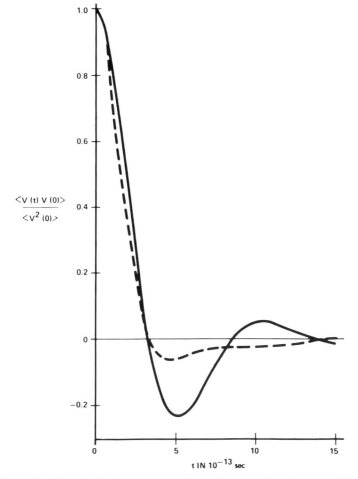

Figure 3.1. Normalized velocity autocorrelation function of liquid argon at 94.4°K. Dashed line is the computer result and the solid line is Equation 3.43 with $\omega_0^2 = 45 \times 10^{24}$ sec^{-2} and $\tau_0 = 1.78 \times 10^{-13}$ sec.

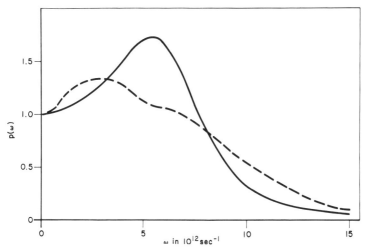

Figure 3.2. Generalized frequency distribution of liquid argon at 94.4°K. Dashed line is computer result and solid line is Equation 3.44.

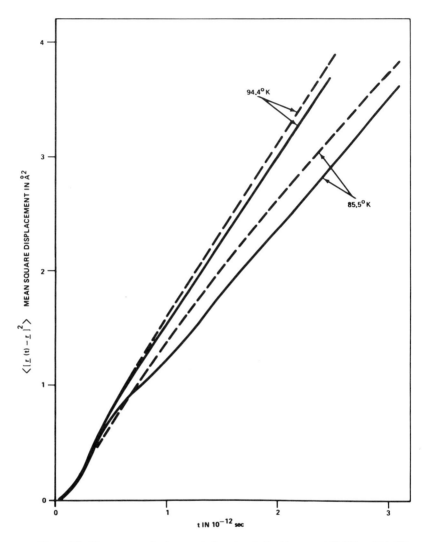

Figure 3.3. Mean square displacement of an atom in liquid argon at 85.5°K and 94.4°K. Dashed lines are computer results and solid lines are Equation 3.45. The self diffusion coefficients (proportional to the slopes at large times) in these cases are 1.88×10^{-5} and 2.43×10^{-5} cm^2/sec, respectively.

Equation 3.45 is used to evaluate the cross section of liquid argon, the results obtained are in quite good agreement with experiment. We can conclude that the model should be adequate for analyzing spectra of molecular liquids, especially when there are only very crude methods for treating rotation effects.

Thus far in our discussions we have mentioned only briefly the effects of molecular rotation in neutron scattering.[20] Although it is widely recognized that $d^2\sigma/d\Omega d\omega$ is quite sensitive to this type of motion, the problem has not received much attention theoretically. The methods that have been developed for evaluating $\chi_R(\kappa, t)$ are limited to two extreme approximations, corresponding to free rotation or small-angle torsional oscillation. As mentioned in Section 2.2, the scattering law for freely rotating rigid molecules of any structure can be computed without approximation. Nevertheless, an approximation method developed by Kreiger and Nelkin has received a good deal of attention.[21] The Kreiger–Nelkin approach is essentially an extension of the Sachs–Teller mass-tensor approximation; strictly speaking, it is useful only if $B \ll \hbar^2\kappa^2/2M \ll k_bT$, where B is a rotational constant. Its popularity apparently stems from the simplifications it introduces in the computation of differential and total cross sections.

If we apply the Gaussian approximation to χ_R, the dynamic problem reduces to the determination of an angular correlation function, $\langle [\mathbf{b}(t) - \mathbf{b}]^2 \rangle$, the analog of the linear displacement function, $\langle [\mathbf{R}(t) - \mathbf{R}]^2$. This approach is intuitively reasonable for a highly hindered system capable of only torsional oscillations. For freely rotating molecules it is in principle not applicable since we already know that χ_R is not a Gaussian function. Nonetheless, one can show that the Gaussian approximation in general leads to a tractable problem, which in the absence of a more microscopic theory gives a reasonably accurate description of χ_R. Even for gases a modified Gaussian approach gives quite good results.[22] We now put

$$\chi_R(\kappa, t) = \exp\left[-\tfrac{1}{2}\kappa^2 \, W_R(t)\right] \qquad (3.46)$$

and

$$
\begin{aligned}
W_R(t) &= \tfrac{1}{3}\langle [\mathbf{b}(t) - \mathbf{b}]^2 \rangle \\
&= \tfrac{2}{3}b^2\left[1 - F_1(t)\right],
\end{aligned} \qquad (3.47)
$$

where

$$F_1(t) = \frac{1}{b^2}\langle \mathbf{b}(t) \cdot \mathbf{b} \rangle. \qquad (3.48)$$

The quantity $F_1(t)$ describes the correlation of a molecular axis at $t = 0$ with itself at time t later. Its time dependence is governed by the molecular reorientation processes characteristic of the system. For isotropic rotations, $F_1(t)$ can be obtained as a Fourier transform of an infrared vibration band.[23] This suggests a method of using optical data directly in the analysis of neutron experiments. Some preliminary results in this direction have been obtained; these will be discussed in the following section.

The problem of determining $F_1(t)$ for a molecular system is considerably simpler than a general study of χ_R. But even in the Gaussian approximation we have no detailed descriptions of the rotational effects. On general grounds we expect $F_1(t)$ to show free-rotation behavior at very short times; at very long times some form of rotational diffusion process should set in. At intermediate times, where intermolecular torque effects are most important, very little is known about the behavior of $F_1(t)$. One can consider the frequency moments of the spectral density of $F_1(t)$ in much the same way as the moments of $p(\omega)$ for the velocity autocorrelation function. Gordon has applied this type of moment analysis to optical bands,[24] but the moments themselves have not yet been used to construct $F_1(t)$ approximately.

In the limit of free rotation and rotational diffusion, one can obtain analytical results for $F_1(t)$. The free-rotation expressions are directly applicable to the analysis of neutron data on molecular gases. Calculations for spherical, linear, and symmetric molecules have recently been carried out.[22, 25] The various rotational correlation functions are

$$F_1(\tau) = \tfrac{1}{3}[1 + 2(1 - \tau^2) \exp(-\tfrac{1}{2}\tau^2)] \qquad \text{(spherical)}, \qquad (3.49)$$

$$F_1(\tau) = \exp(-\tfrac{1}{2}\tau^2) M(-\tfrac{1}{2}, \tfrac{1}{2}, \tfrac{1}{2}\tau^2) \qquad \text{(linear)}, \qquad (3.50)$$

$$F_1(\tau) = \cos^2\phi \, M(-\tfrac{1}{2}, \tfrac{1}{2}, \tfrac{1}{2}\tau^2) \exp(-\tfrac{1}{2}\tau^2) + (1/2 \sin^2\phi) \exp(-\tfrac{1}{2}C\tau^2)$$
$$\times [1 + M(-\tfrac{1}{2}, \tfrac{1}{2}, \tfrac{1}{2}\tau^2) \exp(-\tfrac{1}{2}\tau^2)] \qquad \text{(symmetric)}, \qquad (3.51)$$

where $\tau = t(k_b T/I)^{1/2}$ is a dimensionless time, $c = (I_z - I)/I_z$, ϕ is the angle between **b** and the symmetry axis, and $M(a, b, x)$ is Krummer's confluent hypergeometric function.* These quantities are depicted in Figure 3.4, where as an example of a symmetric molecule we have used the ammonia molecule ($c = 0.365$ and $\phi = 68°$). Notice that for spherical molecules F_1 always remains positive, whereas for linear molecules it can take on negative values. The symmetric molecule can

---

* *Handbook of Mathematical Functions*, edited by M. Abramowitz and I. A. Stegun (Dover Publications, Inc., New York, 1965).

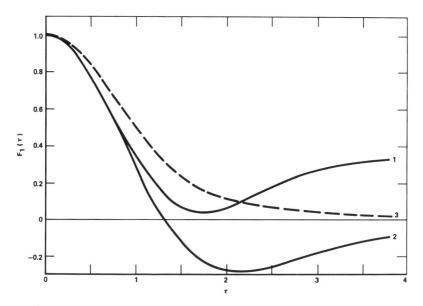

Figure 3.4. The rotational time relaxation function $F_1(\tau)$ for freely rotating spherical (1), linear (2), and symmetric (3) molecules. The symmetric molecule corresponds to ammonia for which $C = 0.365$ and $\phi = 68°$ in Equation 3.51.

exhibit either behavior, depending on the value of ϕ. In fact Equation 3.51 reduces to Equation 3.50 if $\phi = 0$. For a spherical rotator undergoing small-step angular displacements, the rotational correlation function is also known:[26]

$$F_1(\tau) = \exp\{-(2/\xi^2)[\xi\tau + \exp(-\xi\tau) - 1]\}, \tag{3.52}$$

where $\xi = (k_b Tl)^{-1/2}\eta$, η being the rotational friction constant. While this provides a limiting behavior for $F_1(t)$ in condensed systems, Equation 3.52 has not been used explicitly in neutron analysis.

Applications of Equations 3.49 and 3.51 have been made in the case of gaseous CH_4 and NH_3.[22] Combining W_T for an ideal gas (see Section 2.4) and W_R in terms of $F_1(t)$, we can compute the spectra of scattered neutrons following the approach discussed in Section 2.4. (The total width function is just the sum of W_T and W_R.) In Figures 3.5 and 3.6 we show the comparison of experiment with exact quantum-mechanical[27] and Gaussian-approximation results. The rotational transition is clearly seen in both theoretical and experimental spectra, and it is simply a classical effect in molecular reorientation. In the gas

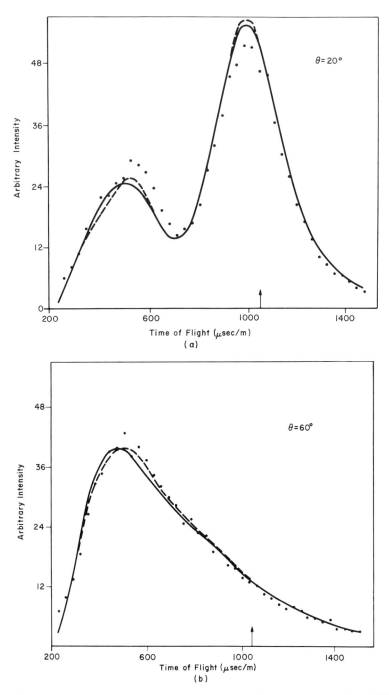

Figure 3.5. Time-of-flight spectra of neutrons scattered by methane gas at $T = 295°$K at various scattering angles θ. Experimental points are given as closed circles, and theoretical results are Gaussian approximation (solid) and exact quantum mechanical (dashed). Calculated spectra have been averaged over an incident spectrum (mean energy 4.87 meV) and are area normalized. (a) $\theta = 20°$, (b) $\theta = 60°$.

62

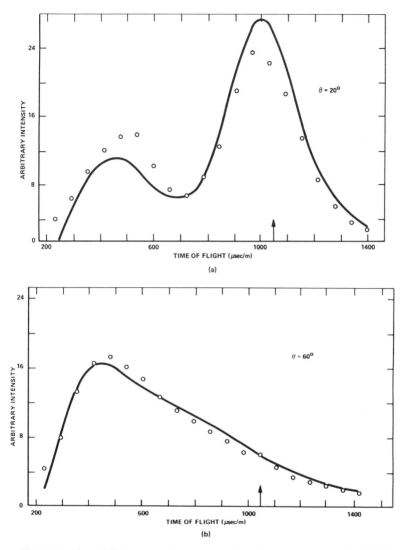

Figure 3.6. Time-of-flight spectra of neutrons scattered by ammonia gas at $T = 295°$K. Notations and symbols same as Figure 3.5.

case it is also possible to study the non-Gaussian effects in χ_R.[22] The leading correction term can easily be included in the calculation, and the results are found to be in excellent agreement with exact calculations. We expect the Gaussian approximation to be even more suitable for liquids and solids. Unfortunately, at the moment there exists no appropriate model of $F_1(t)$ which explicitly takes into account the intermolecular torque effects.

In the absence of a dynamical theory of molecular reorientations, it is not possible to calculate *a priori* the incoherent scattering law for condensed molecular substances. However, within the Gaussian approximation one can use the neutron data to extract a generalized frequency distribution. While this approach, which was discussed in Section 2.4, appears to be quite straightforward, the interpretation of the frequency distribution in the molecular case has not been studied systematically. On the basis of our discussion in Section 3.1, it would appear that the frequency distribution so obtained is a quantity closely related to $G(\omega)$ or $G_H(\omega)$ rather than to $g(\omega)$. On the other hand, from the molecular point of view such a frequency distribution is a composite of translational and rotational contributions, and the question is the relative weight of the two. For example, if the frequencies are to be weighted by associated amplitudes of motions, it would seem that for short times rotational motions are more important than translations. In the following discussion we shall attempt to characterize the respective weights in terms of the molecular mass and an effective rotation mass.[28, 29] This is a very crude concept, but it may provide some physical insight into the problem.

We have already discussed the use of a frequency distribution function $g_T(\omega)$ to describe the center-of-mass motions in neutron scattering. In a similar way a distribution of libration frequencies $g_R(\omega)$ can be introduced through the expression

$$\chi_R(\kappa_1 t) \simeq \exp\left\{-\frac{\kappa^2}{2M_R}\left[\gamma_R(t) - \gamma_R(0)\right]\right\}. \qquad (3.53)$$

where γ_R is defined by Equation 3.38 with g_T replaced by g_R. The only difference between Equation 3.53 and χ_T is the mass factor. In Equation 3.53 M_R is an effective rotation mass that depends on the molecular symmetry. For a linear molecule like HCl, $M_R = 3I/2b^2 = 3m_H M/2(M - m_H)$, where M is the mass of HCl and m_H the hydrogen mass. This result as well as Equation 3.53 can be more or less derived from a hindered rotation model.[30] If there are N hydrogens in the molecule and they are isotropically distributed on the surface of a sphere, and if the nonscattering atoms are assumed to be at the center of the sphere, then $M_R = NM_H$. The intermediate scattering function can now be written as

$$\chi(\kappa, t) = \exp\{-(\kappa^2/2M)[\gamma(t)-\gamma(0)]\}, \tag{3.54}$$

$$\gamma(t) = \int d\omega \, G(\omega)\Lambda(\omega, t), \tag{3.55}$$

$$G(\omega) = g_T(\omega) + (M/M_R) g_R(\omega), \tag{3.56}$$

where both g_T and g_R have unit normalization. The significance of this result is that translation and libration modes do not influence the neutron spectrum to the same extent; that is to say, librational effects are enhanced by a factor of (M/M_R). We have used the symbol $G(\omega)$ to designate the effective frequency distribution in Equations 3.55 and 3.55, thus implying that in the present approach Equation 3.56 is an approximation of Equation 3.24.

Equation 3.54 can be numerically integrated to give the neutron spectrum if $G(\omega)$ is known. If the experiments to be analyzed are carried out under conditions of low sample temperature and small momentum transfer, the one-phonon approximation

$$\chi(\kappa, t) \simeq \exp(-2W)\left[1 + \frac{\kappa^2}{2M}\int_0^\infty d\omega \, G(\omega) \Lambda(\omega, t)\right] \tag{3.57}$$

with $2W = \kappa^2(0)/2M$, can be used. If we ignore the possibility that different hydrogens in the molecule may not be dynamically equivalent, the incoherent-scattering law per hydrogen is the same as Equation 3.26 except that $g(\omega)$ is replaced by $G(\omega)$:

$$S_s^{(1)}(\kappa, \omega) = \exp(-2W)\left[\delta(\omega) + \frac{\hbar\kappa^2 \exp(-\hbar\omega/2k_bT)}{4M\omega \sinh(\hbar\omega/2k_bT)}G(\omega)\right]. \tag{3.58}$$

A measurement of the cross section therefore makes possible an experimental determination of $G(\omega)$. The quantity of more general interest is the thermodynamic frequency function that, in the lattice approach, is the normal-mode frequency distribution $g(\omega)$. In the present treatment it is consistent to regard the low-frequency portion of this function as the sum of translation and libration frequencies:

$$g(\omega) = g_T(\omega) + g_R(\omega). \tag{3.59}$$

We observe that $g(\omega)$ cannot be directly derived from neutron-scattering experiments without additional information. One needs in general an *a priori* knowledge of g_T or g_R; alternatively, under favorable conditions one can assume that g_T and g_R contribute to distinct regions in $g(\omega)$. In a later chapter we apply this approach to ice. Because of the relatively strong intermolecular forces that result from hydrogen bonds and because of the low moments of inertia of the molecule, the derived $g(\omega)$ is at least reliable enough to give thermodynamic predictions in reasonable agreement with measurements.

3.3   Comparison and Correlation with Infrared Spectra

Infrared absorption spectra are conventionally discussed in terms of selection rules and frequency assignment. However, in optical studies of lattice vibrations in solids it is known that valuable information may be derived from absorption overtones and combination bands,[31-33] from impurity-induced absorption spectra,[34] and from Raman scattering.[35] In infrared absorption the elimination of selection rules therefore leads to a more complete investigation of vibration frequencies in molecular substances. Incoherent neutron spectra, by virtue of their band-type character, give information about frequency distributions, and only in certain cases is it possible to obtain precise frequency values for comparison with optical data. In neutron studies multiphonon and anharmonic effects are both undesirable since their presence only complicates the frequency assignment.

Despite the fact that neutron spectra are more similar to the overtones and combination bands in infrared measurements, it is instructive to examine the question of selection rules in these two kinds of experiments. We will not undertake a group-theoretic analysis, but instead will only mention the elementary aspects of this comparison. According to Equation 2.108, a particular transition in the absorption spectrum is governed by the square of matrix elements of $\boldsymbol{\epsilon} \cdot \mathbf{M}$. The matrix element is a measure of the change of $\mathbf{M}_\epsilon$, the component of the total dipole moment operator along $\boldsymbol{\epsilon}$. This change is accompanied by phonon emission because the nuclear displacements in a harmonic crystal can be represented by normal-mode vibrations as in Equations 3.1 or 3.21. Thus $\boldsymbol{\epsilon} \cdot \mathbf{M}$ selects only those phonons whose polarization vector has a nonvanishing component along $\boldsymbol{\epsilon}$. Now $\boldsymbol{\epsilon}$ is perpendicular to the photon wave vector $\mathbf{k}'$, which in turn must be parallel to phonon wave vector $\mathbf{q}$. Therefore, in those cases where one can classify the branches as purely transverse and longitudinal, we see that only transverse phonons can be excited by a fundamental absorption. Furthermore, these have to be optical phonons since the small magnitude of κ implies that acoustic phonons have essentially zero frequencies.

In the neutron case, if the momentum transfer is small, the corresponding coherent and incoherent scattering laws can be written as

$$S(\kappa, \omega) = \delta(\omega) + \frac{1}{nN} \sum_{n'n_0} p(n_0)\, \delta(\omega_{n'} - \omega_{n_0} - \omega)\, |\langle n_0| \sum_{l,\mu} \kappa \cdot \mathbf{R}_l^\mu |n'\rangle|^2, \qquad (3.60)$$

$$S_s(\kappa, \omega) = \delta(\omega) + \frac{1}{nN} \sum_{n'n_0} p(n_0)\, \delta(\omega_{n'} - \omega_{n_0} - \omega) \sum_{l,\mu} |\langle n_0| \kappa \cdot \mathbf{R}_l^\mu |n'\rangle|^2. \qquad (3.61)$$

Comparison of these results with Equation 2.108 shows that κ and ϵ play a similar role in the transition-matrix elements. Evidently both acoustic and optical phonons can be excited in neutron scattering. For one-phonon coherent processes, Equation 3.5 shows that energy and wave-vector conservation conditions must be satisfied,

$$E_f - E_i = \pm \, \omega_j(\mathbf{q}), \qquad (3.62)$$

$$\kappa \pm \mathbf{q} = 2\pi\tau, \qquad (3.63)$$

where upper and lower signs denote phonon absorption and emission, respectively. The polarization vector $\mathbf{A}_j(\mathbf{q})$ and dispersion relation $\omega_j(\mathbf{q})$ are both periodic functions of the reciprocal lattice, so the one-phonon coherent cross section becomes proportional to $|\kappa \cdot \mathbf{A}_j(\kappa)|^2$.[6] By choosing κ along $\mathbf{A}_j(\kappa)$, one can maximize the scattering from the longitudinal branch and suppress any interaction with the transverse mode. The opposite could occur if κ were perpendicular to $\mathbf{A}_j(\kappa)$. This effect, in a certain sense, represents a selection rule that can be externally controlled. For incoherent scattering Equation 3.63 is not required and all branches contribute. Even so, in the case of an oriented sample (for example, stretched polyethylene) longitudinal and transverse vibrations relative to the sample orientation can be picked out by choosing κ perpendicular or parallel to the direction of orientation.

Since the dipole moment operator is linear in the nuclear coordinates, Equation 2.107, in the harmonic approximation, can only describe one-phonon absorption. These transitions appear as sharp lines in the spectrum of a polar crystal, and they correspond to the optically active fundamental modes of vibration. In practice, the observed line shape can be quite broad; in addition, there can occur transitions due to combination bands and overtones.[31, 33] The latter effects are generally indicative of anharmonic forces in the lattice. For diatomic homopolar crystals, the fundamental modes are not infrared-active but are present in the Raman spectrum. However, combination bands can arise in infrared spectra as a result of quadrupole and higher-order interactions [represented by the remaining terms in the expansion of $\exp{(i\mathbf{k}\cdot\mathbf{R})}$ in Equation 2.106]. Whatever their origin, combination bands involve multiphonon processes subject to conservation conditions quite similar to Equations 3.62 and 3.63,[31]

$$\omega = \sum_i \pm \omega_i, \tag{3.64}$$

$$\kappa + 2\pi\tau = \sum_i \pm \mathbf{q}_i, \tag{3.65}$$

where the sum extends over the phonons involved and the upper or lower sign denotes emission or absorption for each phonon. The presence of combination bands implies continuous absorption. However, the spectrum exhibits absorption maxima at critical frequencies corresponding to singularities in the phonon frequency distribution. In this respect a combination band shape may resemble the energy distribution of incoherently scattered neutrons. These are only general statements, and thus far no quantitative correlation between infrared combination bands and neutron spectra has been made.

In contrast to the discrete lines found in gases, infrared rotational spectra of molecular solids and liquids are generally continuous distributions. It is therefore appropriate to investigate molecular reorientation processes in condensed matter in terms of the rotational time correlation function discussed previously. Gordon has shown how $F_1(t)$, the dipole correlation function in this case, and the next-higher-order function can be obtained from the Fourier transforms of infrared and Raman band shapes.[23] Attempts to illustrate the interpretive and computational value of this approach in neutron scattering have been made recently. The first method consists of a partial wave analysis of the neutron cross section in which both infrared and Raman spectra are utilized.[25, 36] The second method involves the Gaussian approximation (see Section 3.2) and makes use of only the infrared data.[37] The two methods are somewhat complementary in that the former gives a more detailed description at small momentum transfers (cold-neutron experiments), whereas the latter should be more applicable in the larger momentum-transfer region (thermal neutrons). The separation of the center-of-mass and rotational motions has to be assumed in both calculations, and as mentioned before it is not clear to what extent this assumption is justified.

In the Gaussian approximation, the mean-square displacement function $W(t)$ is approximately

$$W(t) = \tfrac{1}{3}\langle[\mathbf{R}(t)-\mathbf{R}]^2\rangle + \tfrac{1}{3}\langle[\mathbf{b}(t)-\mathbf{b}]^2\rangle, \tag{3.66}$$

where $\mathbf{R}$ and $\mathbf{b}$ are as defined in Section 3.2. For isotropic rotations, the second term in $W(t)$ can be related to the Fourier transform of a normalized infrared vibration band $I(\omega)$,[37]

$$\tfrac{1}{3}\langle[\mathbf{b}(t)-\mathbf{b}]^2\rangle = \tfrac{2}{3}b^2\left[1 - \int_{-\infty}^{\infty} d\omega\, I(\omega)\exp(-i\omega t)\right]. \tag{3.67}$$

Equation 3.67 establishes a connection between an infrared absorption band and the spectrum of incoherently scattered neutrons. This relation has been used to correlate infrared and neutron measurements of solid and liquid methane.

In Figure 3.7 we show the rotation relaxation function $F_1(t)$ derived[38] from infrared measurement of a vibration band in solid methane at $28°$K.[39] Curves A and B correspond to slightly different ways of smoothing the fluctuations in the experimental $I(\omega)$. For comparison, a free spherical rotator result, Equation 3.49, is also shown. It can be seen that for times less than $t_0 \sim 2 \times 10^{-13}$ sec, there is little difference between $F_1(t)$ (curve A) and the free rotator. The dimensionless time τ_0 is about 0.55, which is essentially the same as that found for liquid methane at $98°$K. Deviation of $F_1(t)$ from free-rotation behavior beyond t_0 indicates the onset of intermolecular torque effects. At long times $F_1(t)$ appears to approach an exponential decay of the form $\exp(-Dt-C)$. The dimensionless rotational diffusion coefficient, $D^* = (I/k_bT)^{1/2}D$ turns out to be around 0.4, about 40 percent smaller than the corresponding value for the liquid. The value of C seems to be unchanged in the solid and liquid phases.

In the present approach the generalized frequency distribution function can be written as

$$p(\omega) = \frac{\hbar\omega/2M}{\sinh(\hbar\omega/2k_bT)}g_T(\omega) + \frac{2Mb^2\omega^2}{3}I(\omega), \tag{3.68}$$

where g_T is the normalized spectrum of translation frequencies. A comparison of Equation 3.67 with the distribution obtained by extrapolation of the neutron data[40] (see Equation 2.88) is shown in Figure 3.8. For g_T a simple Debye spectrum ($\theta_D = 150°$K) was assumed. By treating translations and rotations separately we see that in $p(\omega)$ the rotation effects dominate at practically all frequencies. Moreover, the

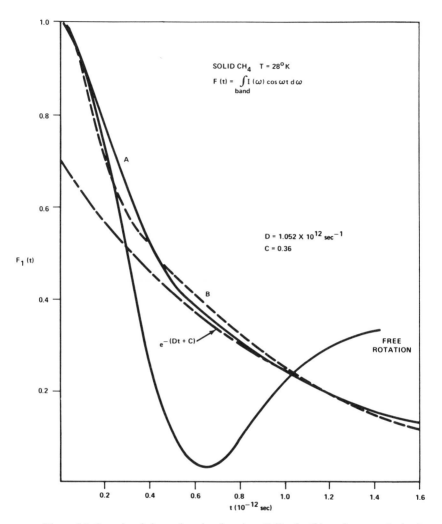

Figure 3.7. Rotational time relaxation function, $F_1(t)$, of solid methane as obtained from an infrared vibration band. Curves A and B correspond to two slightly different ways of smoothing the experimental $I(\omega)$. The free rotation curve is computed from Equation 3.49. The exponential curve denotes a fit at long times which gives a value of 1.052×10^{12} sec^{-1} for the rotational diffusion constant.

rotation contribution in Equation 3.67 is quite different from the free-rotation result. A comparison of calculated and observed neutron spectra shows reasonable agreement except in the vicinity of 7-meV energy transfer.[38] The shoulder in the theoretical cross section arises

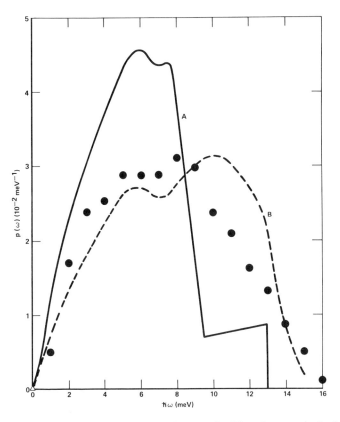

Figure 3.8. The generalized frequency distribution of solid methane as obtained from neutron data at 22.1°K by an extrapolation procedure (closed circles) and as obtained from Equation 3.68 using infrared band shapes (curves A and B correspond to the two $F_1(t)$ shown in Figure 3.7 and a Debye spectrum ($\theta_D = 150°$K) for $g_T(\omega)$.

from a corresponding peak in the rotation part of $p(\omega)$, and this in turn originates from a shoulder in the observed $I(\omega)$.

We can also use Equation 3.68 to illustrate the concept of effective rotation mass discussed in the preceding section. If we rewrite Equation 3.68 formally as

$$p(\omega) = \frac{\hbar\omega/2M}{\sinh (\hbar\omega/2k_bT)} \left[g_T(\omega) + \frac{M}{M_R} g_R(\omega) \right], \qquad (3.69)$$

then by requiring g_R to have unit normalization, we can determine the rotation mass M_R in terms of the second term in Equation 3.68. Using spectrum A of Figure 3.7, we arrive at a value of $M_R \sim 5m_H$, whereas the foregoing qualitative considerations indicate that for methane M_R should be about $4m_H$.[38] Once $p(\omega)$ is expressed in the form of Equation 3.68 and a value for M_R is obtained, an approximate thermodynamic frequency distribution $g(\omega)$ is given by Equation 3.58. Using this method, the value of specific heat c_v computed at 28 °K agrees with the observed value within the quoted experimental uncertainty.

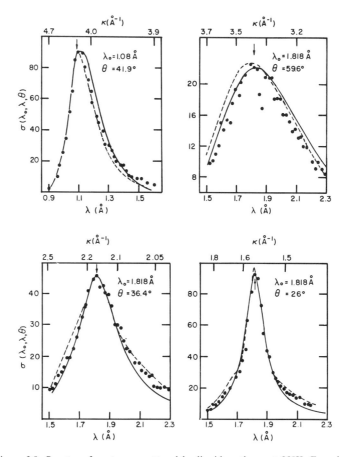

Figure 3.9. Spectra of neutrons scattered by liquid methane at 99°K. Experimental results are given as closed circles, and theoretical results are those derived in the Gaussian approximation (solid lines) and those obtained under the assumption of free rotations (dashed lines). Calculated spectra do not include incident spectrum and resolution effects, and are normalized at the points indicated by an arrow.

Similar calculations of neutron cross sections have been carried out for liquid methane.[37] A comparison with experiment is shown in Figure 3.9, where the dashed lines correspond to a free-rotation quantum-mechanical treatment.[41] In both calculations a Langevin description of the center-of-mass motion was used, but a lower value of the friction constant than that derived from the measured diffusion coefficient was needed to obtain a good fit. This implies that not all the translational degrees of freedom in the liquid are undergoing diffusive motion on the time scales relevant to neutron scattering. Subsequently it has been shown that by using the interpolation model of Section 3.2 essentially the same results can be obtained without distorting the value of D.[42]

REFERENCES

1. *Thermal Neutron Scattering*, edited by P. A. Egelstaff (Academic Press, Inc., New York, 1965).
2. V. F. Turchin, *Slow Neutrons* (Daniel Davey & Company, Inc., New York, 1965).
3. B. N. Brockhouse, in *Phonons and Phonon Interactions,* edited by T. A. Bak (W. A. Benjamin, Inc., New York, 1964).
4. *Inelastic Scattering of Neutrons in Solids and Liquids* (International Atomic Energy Agency, Vienna, 1961, 1963, and 1965).
5. M. Born and K. Huang, *Dynamical Theory of Crystal Lattices* (Oxford University Press, London, 1957).
6. G. Placzek and L. van Hove, *Phys. Rev.* **93**, 1207 (1954).
7. W. M. Lomer and G. G. Low, Chapter 1 in Ref. 1.
8. See G. Dolling and A. D. B. Woods, Chapter 5 in Ref. 1; see also Ref. 3.
9. C. Kittel, *Quantum Theory of Solids* (John Wiley & Sons, Inc., New York, 1963).
10. L. van Hove, *Phys. Rev.* **89**, 1189 (1953).
11. A. Sjolander, *Arkiv Fysik* **14**, 315 (1958).
12. J. Lynch and G. C. Summerfield, to be published. S. Trevino, to be published.
13. See, for example, A. A. Maradudin, and A. E. Fein, *Phys. Rev.* **128**, 2589 (1962). V. Ambegaokar, J. M. Conway, and G. Baym, in *Lattice Dynamics,* edited by R. F. Wallis (Pergamon Press, Inc., New York, 1965).
14. A. Sjolander, Chapter 7 in Ref. 1.
15. P. A. Egelstaff, *Introduction to the Liquid State* (Academic Press, Inc., New York, 1967).
16. B. N. Brockhouse and N. K. Pope, *Phys. Rev. Letters* **3**, 259 (1959); B. N. Brockhouse *et al.,* in Ref. 4, 1963, Vol. I, p. 139; B. A. Dasannacharya and K. R. Rao, *Phys. Rev.* **137**, A417 (1965); K. E. Larsson, Chapter 8 in Ref. 1.

17. A. Rahman, *Phys. Rev.* **136**, A405 (1964); B. R. A. Nijboer and A. Rahman, *Physica* **32**, 415 (1966).
18. R. C. Desai and S. Yip, *Phys. Rev.* **166**, 129 (1968).
19. R. C. Desai and S. Yip, *Phys. Letters* **25A**, 211 (1967).
20. See, for example, J. A. Janik and A. Kowalska, Chapters 9 and 10 in Ref. 1.
21. T. J. Kreiger and M. S. Nelkin, *Phys. Rev.* **106**, 290 (1957).
22. A. K. Agrawal and S. Yip, *Phys. Rev.* **171**, 263 (1968).
23. R. G. Gordon, *J. Chem. Phys.* **43**, 1307 (1965).
24. R. G. Gordon, *J. Chem. Phys.* **39**, 2788 (1963); **40**, 1973 (1964); **41**, 1819 (1964).
25. V. F. Sears, *Can. J. Phys.* **44**, 1279 (1966); **45**, 237 (1967).
26. W. A. Steele, *J. Chem. Phys.* **38**, 2411 (1963).
27. G. Venkataraman, K. R. Rao, B. A. Dasannacharya, and P. K. Dayanidhi, *Proc. Phys. Soc. (London)* **89**, 379 (1966).
28. H. Prask, H. Boutin, and S. Yip, *J. Chem. Phys.* **48**, 3367 (1968).
29. H. Hahn, in Ref. 4, 1965, Vol. 2.
30. S. Yip and R. K. Osborn, *Phys. Rev.* **130**, 1860 (1963).
31. S. S. Mitra and P. J. Gielisse, in *Progress in Infrared Spectroscopy* (Plenum Publishing Corporation, New York, 1964). Vol. II.
32. D. H. Martin, *Advan. Phys.* **14**, 39 (1965).
33. E. Burstein, in *Phonons and Phonon Interactions,* edited by T. A. Bak (W. A. Benjamin, Inc., New York, 1964).
34. A. A. Maradudin, *Solid State Phys.* **19**, 7 (1966).
35. R. Loudon, *Advan. Phys.* **14**, 423 (1965).
36. G. Venkataraman, B. A. Dasannacharya, and K. R. Rao, *Phys. Rev.* **161**, 133 (1967).
37. A. K. Agrawal and S. Yip, *J. Chem. Phys.* **46**, 1999 (1967).
38. A. K. Agrawal and S. Yip, to appear in *Molecular Dynamics and Structure of Solids,* NBS Symposium on Materials Research, 1967.
39. G. E. Ewing, *J. Chem. Phys.* **40**, 179 (1964).
40. Y. D. Harker and R. M. Bragger, *J. Chem. Phys.* **46**, 2201 (1967).
41. G. W. Griffing, *J. Chem. Phys.* **43**, 3328 (1965).
42. A. K. Agrawal, R. C. Desai, and S. Yip, in *Inelastic Scattering of Neutrons in Solids and Liquids* (International Atomic Energy Agency, Vienna, 1968).

Part III
Experimental Studies

4. Polymeric Compounds I. Polyethylene and Related Molecules

The technique of inelastic neutron scattering has been applied to the study of low-frequency motions in several polymers.[1-7] These studies have led to the tentative assignment of frequencies to the various characteristic vibrations of the chain. Such assignments are based on comparisons with observed infrared absorption frequencies, Raman spectra, and various normal-mode calculations.

4.1 Polyethylene

The polymer most intensively studied with this technique has been polyethylene (PE).[2-7] The simplicity of its structure, transplanar, and the fact that it contains only two (CH_2) groups in the repeat unit leads to a relatively small number (9) of normal modes for the chain. Polyethylene is also the only polymer for which a complete normal-mode calculation of the nonfactor group frequencies as a function of δ (phase difference between two adjacent CH_2 units) has thus far been performed.[8-9] These calculations can be used to derive the frequency distribution of the lattice, which can in turn be compared with the frequency distribution derived through an approximation based on neutron-scattering data. The neutron measurements can be used to verify the results of the calculations of the various modes which are in a frequency range now unattainable by optical techniques or which are optically inactive.

According to the model of an infinitely extended isolated molecule,[10, 11] the dispersion relation for polyethylene has only two acoustic branches below the lowest optical branch, which has its lower limit at 720 cm^{-1}. These are the carbon skeletal torsion, with calculated frequency limit near 190 cm^{-1}, and the skeletal stretch-bend mode, with calculated limit near 510 cm^{-1}. The neutron-scattering measurements[1-3] have been interpreted as demonstrating the existence of these two limits. A prominent scattering peak near 195 cm^{-1} is clearly observed in all the low-temperature experiments and appears to verify the torsional-mode limit. The limit for the higher mode, however, has been obscure and is not actually verified.

Measurements of neutron incoherent-scattering cross sections are usually made without regard to the orientation in the lattice of the neutron momentum-transfer vector κ. In particular, all the measurements on polyethylene have utilized targets in which the microcrystalline structure was randomly oriented.

The low-frequency dispersion curves for a polyethylene chain are shown in Figure 4.1. The mode labeled v_5 corresponds to a combination of C–C stretching and C–C–C bending; the mode labeled v_9 corresponds to a torsional vibration around the C–C bond of the polyethylene chain. The dotted lines correspond to splitting of the branches when the effects of the crystalline field are taken into account. Very broad bands have been observed in the infrared spectrum of polyethylene,[12, 13] centered at about 600 and 200 cm^{-1},[12] together with a relatively sharp peak at 71 cm^{-1}.[13] The peaks observed in the neutron spectrum of polyethylene[2] (Marlex 6050) are indicated by circles. Myers *et al.*[7] and S. Trevino[14] have extended these measurements to a polyethylene target which has been given a high degree of uniaxial orientation by stretching. Scattered neutrons were observed by a Brockhouse "constant-Q" technique,[15] applied in this case to incoherent scattering. The method permits discrimination between modes of different phonon polarization so that the response of the longitudinal stretch-bend mode may be enhanced with respect to that of the transverse torsional mode.

The experiments of Trevino[14] and of Myers *et al.*[7] were designed to test the effect of the factor $|\kappa \cdot \mathbf{R}_l|^2$ in Equations 3.21 and 3.60. The momentum transfer κ was placed alternatively parallel and perpendicular to the chain direction. Vibrational modes for which the hydrogen atoms move perpendicular to the chain (v_9) would not be expected to contribute to the scattering when κ is parallel to the chain;

similarly, v_5 should not contribute to the scattering when κ is perpendicular to the chain axis. The peak at 525 cm^{-1} was shown to be a longitudinal mode, and its assignment to v_5 was confirmed.[7] The peak at 200 cm^{-1} was, however, not completely eliminated when κ

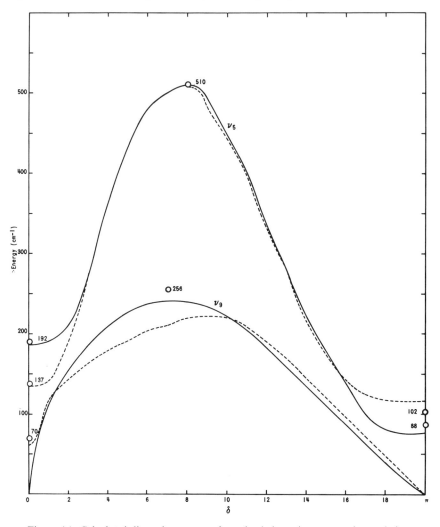

Figure 4.1. Calculated dispersion curves of a polyethylene zigzag transplanar chain taking into account the effects of the crystalline field. The open circles correspond to peaks observed in the time-of-flight spectrum of neutrons inelastically scattered by a polyethylene sample (Marlex 6050).

was placed parallel to the chain. This threw some doubt on its assignment to v_9. Measurement of the neutron scattering from an oriented sample of polyethylene was repeated by Trevino using the time-of-flight spectrometer.[14] The polymer sample was oriented by stretching and rolling. X-ray diffraction measurements show that the polymer chains are aligned parallel to the long direction of the sample with a distribution characterized by an angular width of about 4.7° at half-maximum. The orientation which was used is drawn schematically in Figure 4.2. The term $|\kappa \cdot \mathbf{R}_l^{\prime\prime}|^2$ is given in this figure as a function of energy transfer for the two cases in which transitions are allowed or forbidden, depending on whether the polarization vectors are, respectively, parallel to or perpendicular to κ. The results of the measurement in which the sample temperature was 293°K are shown in Figure 4.3. The time-of-flight spectrum is plotted as a function of energy transfer in cm^{-1}. The general features of the spectra confirm the results of Myers *et al.*[7] The mode v_5 is clearly identified. Again, the peak at 200 cm^{-1} is not completely eliminated by placing κ parallel to the chain. However, there is a definite peak at 240 cm^{-1} in the "perpendicular" measurement which seems to be eliminated in the parallel one. It was therefore suggested that this peak at 240 cm^{-1} be assigned to v_9. Such an assignment is in agreement with a calculation by Tasumi and Schimanouchi[16] of the frequency-versus-phase curves for an extended polyethylene chain in the crystalline field of its neighbors. This calculation shows the effect of the crystalline field on the chain modes and in particular predicts the splitting of v_9 into two rather well-separated branches. The high-frequency limits of the two branches are given as 240 cm^{-1} and 225 cm^{-1}. On the other hand, Kitagawa and Miyazawa[9] have also performed such a calculation. The frequency distribution they obtained is reproduced in Figure 4.4. The high-frequency limits of v_5 and v_9 are given as 570 cm^{-1} and 200 cm^{-1}, and no splitting is shown for v_9. The discrepancy in the two calculations seems to be due to the values of the crystalline force constants used by the different authors, the former authors using a substantially stronger field. The measurements of Trevino[14] seem to favor the results of Tasumi and Schimanouchi although the observation of the splitting of v_9 is not apparent. This may be due in part to the resolution of the instrument used.

In addition to the frequencies associated with chain modes, three other peaks are predicted as shown in Figure 4.4. The broad peak at 140 cm^{-1} is the result of over-all rotatory vibration of the chains about the axis. The peak at 90 cm^{-1} results from antiparallel translatory

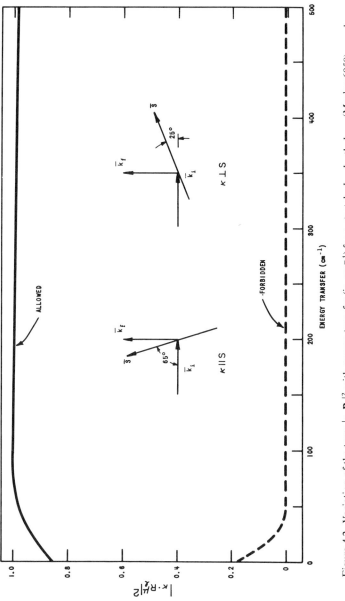

Figure 4.2. Variation of the term $|\boldsymbol{\kappa} \cdot \mathbf{R}_l|^2$ with energy transfer (in cm^{-1}) for a stretched polyethylene (Marlex 6050) sample.

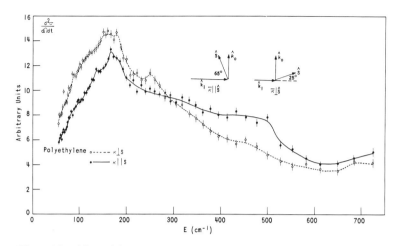

Figure 4.3. Differential neutron cross section plotted as a function of energy transfer (in cm^{-1}) of neutrons inelastically scattered at a 90° angle from an oriented sample of polyethylene at 293°K. ○ , $\kappa \perp \mathbf{S}$ and ● , $\kappa \| \mathbf{S}$ where κ is the momentum transfer $(\mathbf{k}_i - \mathbf{k}_f)$ of the neutron and $\mathbf{S}$ the direction of orientation of the chains.

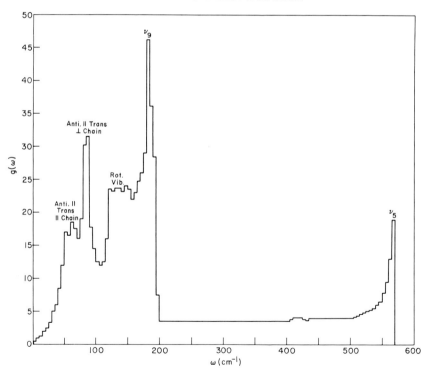

Figure 4.4. Calculated frequency distribution of an "ideal" sample of polyethylene.

82

vibrations of adjacent chains perpendicular to the chain axis. Finally, the peak near 60 cm^{-1} is the result of antiparallel translatory vibrations parallel to the chain axis. The time-of-flight spectrum obtained by Danner et al.[2] shows a peak at 240 cm^{-1} for which no interpretation was provided. If one assigns this peak to v_9, there are three peaks below v_9 which could be assigned to the three lattice peaks of Kitagawa and Miyazawa.[9] There are several possible explanations for the additional broad band centered at 325 cm^{-1} in the neutron spectrum: (a) The sample of polyethylene used in this measurement was high-density Marlex 6050. The sample certainly contained regions of disorder and defects. The influence of these regions on the motions of the ideal extended chain in the well-ordered crystalline field of its neighbors could possibly give rise to some of the unexplained peaks. (b) These peaks could correspond to combination bands that can be observed in the neutron spectra as in the infrared spectra, although they are expected to be very weak. (c) Possibly the most reasonable explanation is that the approximate expression (Equation 3.58) used to obtain $G(\omega)$ from the data constitutes a rather poor approximation. The effect of the amplitude of vibration on the cross section might produce the discrepancy in the relative heights of the observed structure. Only an exact calculation of the cross section can detect the size of this effect. Recent calculations by Summerfield[17] have shown that even at a temperature of 100°K multiphonon scattering effects contribute appreciably to the scattered neutron spectrum and are probably responsible for the broad band observed by Danner et al. at about 325 cm^{-1}.

The specific heat of crystalline polyethylene has been determined by Wunderlich.[18] The calculation of this quantity from any proposed frequency distribution should constitute a test of its validity. Kitagawa and Miyazawa[19] calculated the specific heat up to 150°K using their frequency distribution and obtained quite a good fit. If the specific heat is calculated using the frequency distribution $G(\omega)$, the agreement is rather poor. If, on the other hand, the relative height of the various peaks is chosen so as to conform with the calculation of Kitagawa and Miyazawa but their energies are chosen from the neutron-scattering data, one obtains the frequency distribution shown in Figure 4.5, from which a specific heat can be calculated in somewhat better agreement with the measured specific heat.

In summary, there is some evidence from measurements on the oriented sample of polyethylene that the high-frequency limit of v_9

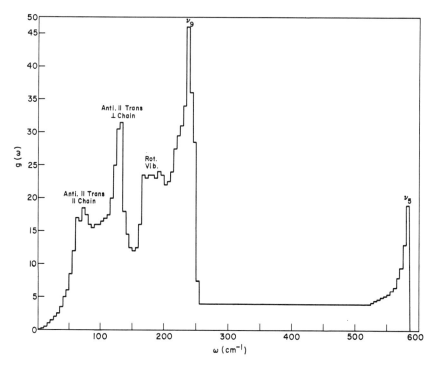

Figure 4.5. Frequency distribution of polyethylene derived using the neutron data and the calculation of Kitagawa and Miyazawa.[9]

could occur at 240 cm^{-1}; the three peaks below this would then be assigned to the lattice modes. These frequencies are in reasonable agreement with those predicted by tne latest calculation of Tasumi and Schimanouchi.[16]

4.2 Polypropylene

Investigations similar to those on polyethylene have been made on polypropylene to determine the low-frequency vibrational motions between 1000 and 30 cm^{-1}. Samples ranging from highly isotactic to completely atactic have been studied by various techniques.[20-28] Molecular motions in the highly isotactic and the completely atactic samples have also been measured as a function of temperature, with emphasis on the changes observed in these motions above and below both the melting point and the glass transition.

Due to the asymmetry of the alternate carbon atoms, linear, polypropylene can adopt three steric modifications—syndiotactic, isotactic, and atactic—in which the optical forms of the successive asymmetric carbon atoms exhibit alternate, identical, and random steric configurations, respectively. Initial X-ray studies[20] revealed that the individual chains in crystalline polypropylene were isotactic, possessing a helical configuration. Subsequently, a second crystalline form of isotactic polypropylene was observed.[21] However, the infrared spectra of these two forms were essentially identical, indicating that intermolecular forces are of secondary importance in determining the frequencies of the normal vibrations in isotactic polypropylene.

The optically active phases (corresponding to $\delta = 0$ and $2\pi/3$) of the fundamental vibrations of an infinite isotactic polypropylene chain in its helical configuration have been calculated by Miyazawa and his co-workers.[22] According to their calculation, which neglects intermolecular forces, the vibrations below 600 cm^{-1} arise from skeletal modes involving the deformations of the CH_2–$CH(CH_3)$–CH_2 and CH–CH_2–CH groups and torsional oscillations about both the axial and the equatorial C–C and C-methyl bonds. These modes are not localized in a single repeating unit and are expected to give rise to well-defined modes only for the highly ordered helical structure of isotactic polypropylene. The modes between 700 and 1200 cm^{-1} are mixed combinations of CH_3 and CH_2 rocking modes and C–C stretching modes. The high-frequency optical modes above 1200 cm^{-1} lie well outside the range of the neutron experiment. Agreement was obtained, however, between the predicted frequencies and the observed fundamental bands in the infrared spectra.[22–24]

While fewer systematic optical studies have been made of the molecular motions in atactic polypropylene, results indicate that the chains depart strongly from the long uniform helical configuration found in the isotactic material. The considerable simplification observed in the infrared spectrum of atactic polypropylene[22–24] is attributed to the fact that chains are no longer in the uniform helical configuration, although a recent study of the stability of a helical molecular configuration[25] suggests that even in such noncrystalline material a helical conformation may be maintained in short sections of the chain. It has been postulated[22] that the skeletal deformation and torsional modes lying below 600 cm^{-1} would be sensitive to both an increase in atacticity and an increase in the noncrystalline region of the sample, giving rise to new bands in this frequency region.

Internal rotations, which involve transitions from one local equilibrium configuration to another, have been observed in partially crystalline polypropylene by nuclear magnetic resonance line-width measurements[26, 27] over the temperature range $77°K-435°K$. The initial line narrowing observed between $77°K$ and $130°K$ has been associated with the onset of the methyl group rotation in the crystalline portions of the sample. The appearance and further narrowing of a sharp component between $220°K$ and $300°K$ has been attributed to segmental motion in the amorphous regions involving short and long sections of the chain, respectively. These conclusions are supported by dynamic-mechanical measurements.[27] At lower temperatures such segmental reorientations take place with an average relocation time of 10^{-5} sec; they are not observed in the neutron experiment since the neutron interaction time is of the order of 10^{-12} sec. However, as the sample temperature is increased, internal rotations become less hindered. Although the hindering potentials are not accurately known for semi-crystalline polypropylene, the fact that CH_3 rotation sets in well below $293°K$ suggests that these rotations may be only weakly hindered at room temperature.

The frequencies observed in the neutron spectra of PP-1[28] are compared in Table 4.1 with the calculated optically active frequencies of the fundamental modes for an isotactic helical polypropylene chain[22] and the bands observed in the infrared spectra.[22, 23] Inspection of this table shows that there is reasonable agreement between the calculated frequencies and the results of infrared and neutron measurements. The observed neutron bands are listed opposite the fundamental modes for which there exists the closest numerical agreement between observed and calculated frequencies. It should be emphasized, however, that this assignment may not be correct in all cases. In the neutron measurements, which are sensitive to the entire frequency distribution of a vibrational mode, the maximum of a given mode may occur at a frequency other than an optically active frequency if its dispersion curve possesses a maximum or minimum at an intermediate value of δ. Thus, an unambiguous comparison of the derived frequency distributions with theory requires detailed knowledge of complete dispersion curves of the fundamental modes. Since these curves have not been calculated, the assignments suggested for the individual neutron bands observed in isotactic polypropylene, as discussed below, are tentative in some cases.

At $T = 114°K$ and $T = 293°K$, a peak appears at 100 cm^{-1} which

Table 4.1. Fundamental Vibrational Modes for an Isotactic Helical Polypropylene Chain: Comparison of Calculated Optically Active Frequencies with Observed Bands in Infrared and Neutron Spectra

Mode	Assignments	Frequencies (cm^{-1}) Calculated[a]		Infrared[a]		Neutron
		$\delta = 0$	$\delta = 2\pi/3$	$\delta = 0$	$\delta = 2\pi/3$	
ν_{26}	Torsional oscillation about axial C—C bond.	0	70	...	106	100
ν_{25}	CH—CH$_2$—CH bending; asymmetric CH$_2$—CH(CH$_3$)—CH$_2$ deformation; torsional oscillation about equatorial C—C bond.	138	145	155	169	180
ν_{24}	Torsional oscillation about C-methyl bond.	126	222	200	210	...
ν_{23}	Asymmetric CH$_2$—CH(CH$_3$)—CH$_2$ deformation; symmetric CH$_2$—CH(CH$_3$)—CH$_2$ deformation; torsional oscillation about C-methyl bond.	266	294	251	321	250 315
ν_{22}	Symmetric CH$_2$—CH(CH$_3$)—CH$_2$ deformation; asymmetric CH$_2$—CH(CH$_3$)—CH$_2$ deformation.	381	448	398	460	470
ν_{21}	Asymmetric CH$_2$—CH(CH$_3$)—CH$_2$ deformation; symmetric CH$_2$—CH(CH$_3$)—CH$_2$ deformation; CH—CH$_2$—CH bending.	445	491	456	528	570
ν_{20}	CH$_2$ rocking; equatorial C—C stretching; C-methyl stretching.	842	803	854	809	840

[a] δ = phase difference between adjacent monomer units.

agrees with the frequency calculated for the torsional oscillation mode about the axial C–C bonds; it disappears rapidly with both increasing temperature and increasing atacticity. This is an acoustic mode—consequently, the high-frequency limit lies slightly higher than the peak frequency (that is, at 110 ± 10 cm^{-1})—analogous to the torsional acoustic mode observed in highly crystalline polyethylene.[2] Since axial torsion is a cooperative vibration of many monomer units, the decrease in the definition of this motion with increasing atacticity is to be expected.

Although nuclear magnetic resonance[26, 27] and dynamic-mechanical measurements[27] indicate the onset of CH$_3$ rotation in semicrystalline polypropylene at 77°K, only torsional oscillation of the methyl groups is expected to be observed at low temperature in the neutron experiment as a consequence of the short neutron interaction time (10^{-12} sec) compared with the average reorientation time (10^{-5}–10^{-6} sec) of these groups. The active phases of this torsional mode are predicted by theory and observed in the infrared spectra at about 210 cm^{-1}. The closest neutron band is the shoulder at 180 cm^{-1} in the frequency spectrum of PP-1 at 114°K.[28] Although this band is associated with the mode v_{25}, it may contain relatively weak, unresolved contributions from v_{24}. As the temperature is increased and the CH$_3$ groups rotate more and more freely, the probability of observing the torsional oscillations of the groups diminishes; thus the absence of this mode in the frequency distributions at room and higher temperatures is not unexpected. The pronounced decrease in the intensity of the elastic peak with increasing temperature is attributed in part to an increase in multiphonon scattering as the CH$_3$ groups become more freely rotating.

As shown in Table 4.1, the remaining fundamental modes below 600 cm^{-1} are, in general, combinations of skeletal-deformation normal modes. Of the five neutron bands observed at 570, 470, 315, 250, and 180 cm^{-1} in the frequency distribution of PP-1 at low temperature, the bands at 470, 250, and 180 cm^{-1} correlate reasonably well with the modes v_{22}, v_{23}, and v_{25}, respectively.

The vibration spectra of atactic polypropylene exhibit considerably less well-defined structure than is observed for isotactic polypropylene. Below the glass transition, weak bands are observed at 480 cm^{-1} and 250 cm^{-1} which correlate with the skeletal optical modes v_{22} and v_{23} characteristic of the isotactic material. Their appearance suggests that the CH$_2$–CH(CH$_3$)–CH$_2$ deformation modes are still defined in the atactic polymer. A line at 251 cm^{-1} has been observed in the infrared spectrum of atactic polypropylene.[22] However, the most pronounced feature of the low-temperature frequency distribution is the peak at 750 cm^{-1}. Although shifted to higher frequencies, the shape of this peak is similar to a band at 570 cm^{-1} in the frequency distribution of isotactic polypropylene, which band has been assigned to the high-frequency portion of the acoustic deformation mode. In view of the rigidity of the atactic chain below the glass transition, an acoustic deformation mode is to be expected. It seems plausible, therefore, to associate this peak with such a mode in which the lower frequencies (that is, longer wavelengths) are attenuated as a result of their inability to propagate along the disordered atactic chain. The fact that the temperature dependence of this mode is similar to that observed for the corresponding band at 570 cm^{-1} in PP-1, as well as to that observed for the high-frequency portion of the acoustic deformation mode in polyethylene, supports this hypothesis.

Above the glass transition, the frequency distribution of atactic polypropylene shows a number of weak bands. It has been suggested[22] that since skeletal deformation frequencies, as well as torsional frequencies, have been observed to be sensitive to changes in the skeletal structure, atactic polypropylene may be expected to show new bands arising from skeletal deformation modes in the low-frequency region. While the exact origin of these bands in the atactic polymer is uncertain, the fact that the frequencies of several (those at 480 and 250 cm^{-1}) again correlate with the skeletal optical modes v_{22} and v_{23} characteristic of the isotactic polymer suggests that the CH_2–$CH(CH_3)$–CH_2 deformation vibrations may be the least sensitive to the degree of tacticity and may preserve their identity for a nonhelical chain in the atactic material.

4.3 Polyacrylonitrile

The conformation and tactic structure of polyacrylonitrile (PAN) have not been unambiguously determined because of the generally diffuse X-ray diffraction patterns. Bohn *et al.*[29] have interpreted these patterns as indicating the lack of any valid repeat distance along the chain. There is general agreement about a lateral packing of cylinders within which the chain exists but no evidence that the chains assume any specific structure in these cylinders. Heating the sample changes the lateral order from hexagonal to orthorhombic. There are at least two reported transition temperatures in PAN, 87°C and 140°C. Andrews and Kimmel[30] have studied the 140°C transition with birefringence and show that an irreversible ordering occurs when the polymer is heated above this temperature and allowed to cool. They attribute this to the loosening and reformation of a dipole association structure. Bohn *et al.* have also noted this ordering upon heating in their X-ray measurements. There is, however, no reported increase in the longitudinal order upon heating.

The infrared spectra of PAN have been obtained by several investigators[31, 32] and have been analyzed with some degree of success using a transplanar model for the chain. There are differences in the observed frequencies and, since different tactic structures were assumed, differences in the assignments of various bands. Yamadera *et al.*[32] conclude from the results of infrared and normal-coordinate calculation that PAN may contain a structure very like the syndiotactic planar zigzag, though enough discrepancies exist so that deviation from this structure cannot be eliminated.

The spectra of neutrons inelastically scattered from two samples of PAN have been obtained by Trevino and Boutin.[33] Sample 1 was a fine-grain powder. Its X-ray diffraction pattern showed a large, rather sharp reflection corresponding to a spacing of approximately 5.2Å, a weaker reflection corresponding to a spacing of approximately 3Å, and a general diffuse scattering over a large portion of the pattern. Sample 2 was a film cast from solution (H_2O and NaSCN). Its X-ray diffraction pattern also showed the 5.2Å reflection but considerably weaker and much broader than that of Sample 1. The 3Å reflection was almost completely lost in the greatly increased diffuse scattering. Upon heating above 150°C, however, the X-ray spectrum of Sample 2 sharpened and became similar to that of Sample 1. The frequency distributions $[G(\omega)]$ derived from the neutron measurements are shown in Figure 4.6. The peaks centered at 560 cm^{-1} and 265 cm^{-1} agree with infrared lines observed in these energy regions and can be assigned to low-lying optical modes (bending or wagging). The curve displays a minimum at about 430 cm^{-1} where an infrared line has been reported.[31] These two observations, however, are not necessarily in disagreement. Some of the optically active modes need not occur in regions where there exists a large density of states. The broad shoulder extending from approximately 50 to 160 cm^{-1} covers the frequency range in which the low-energy acoustic modes are expected. Any specific structure in this part of the spectrum is taken as an indication of the degree of order along the chain, since these low-energy acoustic modes are expected to be most sensitive to chain order and should be

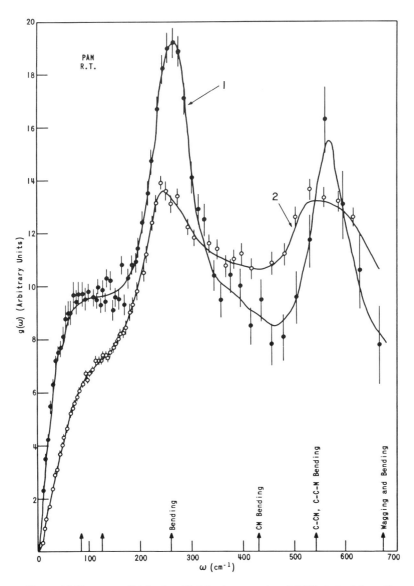

Figure 4.6 Frequency distribution $G(\omega)$ for two samples of PAN, derived from the neutron measurements.

the first to be broadened upon disordering of the chain. The differences in the neutron spectra of the two samples are striking. The two bands associated with the optical modes are sharper, and the low-energy shoulder is better defined in Sample 1 than in Sample 2. The relatively high degree of order observed in Sample 1 reflects polymerization directly to a powder without passing through a solution or melt. The fact that the neutron spectrum of Sample 2 becomes similar to that of Sample 1 after a temperature cycling up to 150°C indicates the existence of chain ordering which increases irreversibly as the temperature is raised through the transitions. This chain ordering, which had not been suspected from X-ray studies, may be a consequence of the better packing of the chains. Since no change in tacticity is expected to take place with increasing temperature, the ordering along the chain probably corresponds to an increase in length of the segments of the chain in a transplanar configuration.

4.4 Polyoxymethylene

The molecule of polyoxymethylene (POM) assumes a helical conformation and contains nine monomer units (CH_2O) per identity period.[34] Infrared and Raman spectra of this polymer have been obtained for certain cases in order to make band assignments. Several fundamental vibrations of the isolated chain are expected to occur below 200 cm^{-1}.

In a study undertaken by Trevino and Boutin,[33, 35] the neutron spectrum of a commercial sample of POM in powder form was obtained at room temperature. The frequency distribution $G(\omega)$ derived from this measurement is shown in Figure 4.7. The structure below 200 cm^{-1} is due to the acoustic modes of the chains which are optically inactive. The three lowest vibration frequencies corresponding to optical modes which have been observed by infrared[35] are approximately 237, 483, and 634 cm^{-1}.

In order to attempt an assignment of the bands observed in the neutron spectrum, additional measurements were obtained from an oriented sample of POM.[35] In this case, the momentum transfer κ of the scattered neutron was placed alternately parallel and perpendicular to the chain axis. The resulting frequency distributions reflect, as was shown in the case of oriented polyethylene,[7, 14] vibrations in which hydrogen atoms move along or move transverse to the chain axis, respectively. The frequency distributions obtained for POM are shown

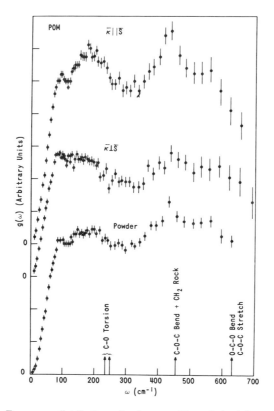

Figure 4.7. Frequency distribution of polyoxymethlene derived from the neutron data. The lower curve corresponds to a polycrystalline sample. The two upper curves correspond to a stretched oriented sample of POM such that the momentum transfer κ of the neutron is alternately $\parallel$ or $\perp$ to the direction of orientation of the chain.

in Figure 4.7. The peaks at 430 and 180 cm^{-1} correspond to longitudinal motions since their intensities decrease appreciably when κ is perpendicular to the direction of the chain axis. The peak at 180 cm^{-1} is assigned to a mode analogous to mode ν_5 (C–C–C bending and stretching) in polyethylene.[2, 14] The peak at 430 cm^{-1} also observed in the infrared spectrum is due to C–O–C bending mixed with CH$_2$ rocking. The peak around 80 cm^{-1} in the neutron spectrum is observed with both polarizations. It may correspond to a mode similar to mode ν_9 (C–C torsion) in polyethylene. A shoulder is also observed in the neutron spectrum around 130 cm^{-1}. The infrared frequency at 235 cm^{-1}

has been assigned to a torsional vibration.[36] The neutron spectrum in that frequency region shows only a broad distribution which is not well resolved from the 180 cm^{-1} band. There is also evidence of some structure in the 500–600 cm^{-1} range which might correspond to infrared-inactive frequencies.

Along with infrared results, these measurements can provide a useful basis for calculating the frequency distribution of the vibrations of a POM chain, a task that has not yet been attempted.

4.5   Nylon-6

Vibration spectra in the 1500–8 cm^{-1} range have been measured for a sample of high-density, fiber-grade Nylon-6 by the inelastic scattering of slow neutrons.[37] The frequency region above 700 cm^{-1} has been extensively investigated by infrared studies.[38, 39] Results of infrared and neutron measurements are compared in Table 4.2. The intramolecular cooperative vibrations of the $(CH_2)_5$ groups are expected to occur below 600 cm^{-1} and are compared with the neutron

Table 4.2. Observed Vibration Frequencies in Nylon-6

Assignments	Infrared Frequencies (cm^{-1})	Neutron Scattering Frequencies (cm^{-1})	
NH stretching	3300–3070[a]	2944 ± 70[d]	
NH deformation	1545[a]		
NH deformation	1280[a]	1500 ± 100[d]	
CH$_2$ bend	1420–1480[b]		
CH$_2$ twist	1200–1370[b]		
CH$_2$ rocking	650–800[a–c]	800 ± 40[d]	
Torsional vibration about CO—NH bond	217[c]	184 − 204[d]	
n-Heptane modes (five CH$_2$ units)		148 ± 8	316 ± 15[d, e]
		117 ± 7	260 ± 9
			194 ± 8
		86 ± 6	168 ± 8
		73 ± 5	43 ± 3

[a] C. G. Cannon, *Spectrochim. Acta* **16**, 302 (1960).
[b] M. L. Tobin and M. J. Carrano, *J. Chem. Phys.* **25**, 1044 (1956).
[c] T. Miyawaza, *J. Chem. Phys.* **32**, 1647 (1960).
[d] G. J. Safford and F. L. LoSacco, *J. Chem. Phys.* **43**, 3404 (1965).
[e] H. Boutin, H. Prask, S. F. Trevino, and H. R. Danner, in *Inelastic Scattering of Neutrons in Solids and Liquids* (International Atomic Energy Agency, Vienna, 1965), Vol. 2, p. 407.

spectra reported previously for solid n-heptane.[5] The degree of agreement between the peaks observed in Nylon-6 and in solid n-heptane suggests a similarity of the lattice modes involving the five CH_2 units in the two materials. In the n-paraffins it has been shown that the influence of intermolecular forces on the intramolecular modes may be neglected with a relatively high degree of accuracy.[40] In Nylon-6, the packing of the chain segments is much more open than in the n-paraffins, and a certain amount of vibration or twisting of the $(CH_2)_5$ segments can probably occur without unduly disturbing the chain packing or the amide-amide contacts.[39] It is expected from previous infrared measurements and from theoretical considerations[40] that the terminal-group vibrations should influence the intramolecular frequencies of short chain segments. However, the similarity of the end-group torsional frequencies (217 cm^{-1} for NH–CO torsion[41] and 211 cm^{-1} for CH_2–CH_3 torsion in n-heptane[42]) may explain the observed similarities in the vibrations of the $(CH_2)_5$ segments of these compounds.

REFERENCES

1. G. J. Safford, H. R. Danner, H. Boutin, and M. Berger, *J. Chem. Phys.* **40**, 1426 (1964).
2. H. R. Danner, G. J. Safford, H. Boutin, and M. Berger, *J. Chem. Phys.* **40**, 1417 (1964).
3. J. S. King and J. L. Donovan, *Bull. Amer. Phys. Soc.* **9**, 623 (1964).
4. W. L. Whittemore, "Scattering of Neutrons by Polyethylene," Rept. GA–6456, Gulf General Atomic, La Jolla, Calif. (1965).
5. H. Boutin, H. Prask, S. F. Trevino, and H. R. Danner, "Inelastic Scattering of Neutrons," in *Inelastic Scattering of Neutrons in Solids and Liquids* (International Atomic Energy Agency, Vienna, 1965), Vol. 2, p. 407.
6. W. Myers, J. L. Donovan, and J. S. King, *J. Chem. Phys.* **42**, 4299 (1965).
7. W. Myers, G. C. Summerfield, and J. S. King, *J. Chem. Phys.* **44**, 184 (1966).
8. M. Tasumi and T. Schimanouchi, *J. Chem. Phys.* **43**, 1245 (1965).
9. T. Kitagawa and T. Miyazawa, *Rept. Prog. Polymer Phys. (Japan)* **8**, 53 (1965).
10. T. P. Lin and J. L. Koenig, *J. Mol. Spectry.* **9**, 228 (1962).
11. M. Tasumi, T. Shimanouchi, and T. Miyazawa, *J. Mol. Spectry.* **9**, 261 (1962).
12. S. Krimm, C. Y. Liang, and G. B. B. M. Sutherland, *J. Chem. Phys.* **25**, 543 (1956).
13. S. Krimm and M. I. Bank, *J. Chem. Phys.* **42**, 4059 (1965).
14. S. Trevino, *J. Chem. Phys.* **45**, 757 (1966).
15. B. N. Brockhouse, in *Inelastic Scattering of Neutrons in Solids and Liquids* (International Atomic Energy Agency, Vienna, 1961), p. 113.
16. M. Tasumi and T. Schimanouchi, *J. Chem. Phys.* **43**, 1245 (1965).
17. G. C. Summerfield, *J. Chem. Phys.* **43**, 1079 (1965).

18. B. Wunderlich, *J. Chem. Phys.* **37**, 1203 (1962).
19. T. Kitagawa and T. Miyazawa, *J. Chem. Phys.* **47**, 337 (1965).
20. G. Natta, *J. Polymer Sci.* **16**, 143 (1955).
21. G. Natta, M. Peraldo, and P. Corradini, *Rend. Accad. Nazl. Lincei* **8**, 26 (1959).
22. T. Miyazawa, Y. Ideguchi, and K. Fukushima, *J. Chem. Phys.* **38**, 2709 (1963). T. Miyazawa, K. Fukushima, and Y. Ideguchi, *J. Polymer Sci.* **B1**, 385 (1963).
23. M. Peraldo and M. Farina, *Chim. Ind. (Milan)* **42**, 1349 (1960).
24. C. Y. Liang and F. G. Pearson, *J. Mol. Spectry.* **5**, 290 (1960). C. Y. Liang, M. R. Lytton, and C. J. Boone, *J. Polymer Sci.* **47**, 139 (1960); **54**, 523 (1961).
25. P. De Santis, E. Giglio, A. M. Liquori, and A. Ripamonti, *J. Polymer Sci.* **A1**, 1383 (1963).
26. W. P. Slichter and E. R. Mandell, *J. Appl. Phys.,* **29**, 1438 (1958).
27. J. A. Sauer and A. E. Woodward, *Rev. Mod. Phys.* **32**, 88 (1960).
28. G. J. Safford, H. R. Danner, H. Boutin, and M. Berger, *J. Chem. Phys.* **40**, 1426 (1964).
29. C. R. Bohn, J. R. Schaefgen, and W. O. Statton, *J. Polymer Sci.* **55**, 531 (1961).
30. R. D. Andrews and R. M. Kimmel, *J. Appl. Phys.* **35**, 3194 (1964). R. D. Andrews and R. M. Kimmel, *Polymer Letters* **3**, 167 (1965).
31. C. Y. Liang and S. Krimm, *J. Polymer Sci.* **31**, 513 (1958).
32. R. Yamadera, H. Tadokoro, and S. Murahashi, *J. Chem. Phys.* **41**, 1233 (1964). T. Miyazawa, *J. Chem. Phys.* **35**, 693 (1961).
33. S. Trevino and H. Boutin, *J. Macromol. Sci. (Chem.)* **A1**(4), 723 (1967).
34. H. Tadokoro, T. Yasumoto, S. Murahashi, and I. Nitta, *J. Polymer Sci.* **44**, 266 (1960).
35. S. Trevino and H. Boutin, *J. Chem. Phys.* **45**, 2700 (1966).
36. H. Tadokoro, M. Kobayashi, Y. Kawaguchi, A. Kobayashi, and S. Murahashi, *J. Chem. Phys.* **38**, 703 (1963) and references therein.
37. G. J. Safford and F. L. LoSacco, *J. Chem. Phys.,* **43**, 3404 (1965).
38. M. C. Tobin and M. J. Carrano, *J. Chem. Phys.* **25**, 1044 (1956).
39. C. G. Cannon, *Spectrochim. Acta* **16**, 302 (1960).
40. H. Matsuda, K. Okada, T. Takasa, and T. Yamamoto, *J. Chem. Phys.* **41**, 1527 (1964).
41. T. Miyazawa, *J. Chem. Phys.* **32**, 1647 (1960).
42. J. H. Schachtschneider and R. G. Snyder, *Spectrochim. Acta* **19**, 117 (1963).

5. Polymeric Compounds II. Polyglycine and Related Molecules

5.1 Polyglycine

Neutron-scattering data for the polyglycine molecule is discussed below in some detail for several reasons: (a) The normal-mode calculation has been performed for the molecule in the transplanar conformation, and dispersion curves have been derived which can be used in the interpretation of the neutron data. (b) There are a number of infrared studies on polyglycine, the results of which can be compared with the neutron results. (c) Neutron-scattering measurements have been performed for two reasonably well-defined conformations: transplanar and helical. (d) We have here a good illustration of the type of information neutron studies can provide and an indication of their potential importance for investigation of more complex molecules or systems of biological interest.

Polyglycine $(-CH_2CO-NH-)_n$ is the simplest of all amino acid polymers; a study of its normal modes and their dispersion is essential to an understanding of the more complex polypeptides and proteins. It exists in two conformational forms. X-ray diffraction studies[1] show that polyglycine I is the β form in which the fully extended chains have a twofold screw axis and are packed in an antiparallel arrangement. Neighboring chains are hydrogen-bonded via N—H---O—C. Polyglycine II chains exist in helical form with threefold screw axes. The chains are primarily parallel and are packed in a hexagonal lattice.[2] The arrangement in which all chains have their peptide sequence in the same direction appears to be the simplest. However, the simultaneous presence of both direct and inverted chains in polyglycine II has been noted by Padden and Keith[3] in studies of crystalline morphology and by Krimm[4] in infrared studies. In a recent communication, Ramachandran et al.[5] have shown that, as in the structure of collagen,[6] hydrogen bonds of the type C—H---O can be formed in polyglycine II. They have also shown that N—H---O hydrogen bonds of approximately the same strength can occur not only between chains running in the same direction but also between chains running in opposite directions. However, C—H---O hydrogen bonds are formed between like chains and are very weak between chains running in opposite directions. These observations are in accord with the infrared studies of Krimm, Kuroiwa, and Rebane.[7] Ramachandran et al.[8] have also shown that in polyglycine II only one-third of the possible C—H---O bonds are formed.

Recently Fukushima et al.[9] have made a normal-mode calculation limited to the in-plane vibrations of polyglycine I and have studied the amide band frequencies. Since the low-frequency vibrations are expected to depend sensitively on the chain conformation, normal modes for both the in-plane and the out-of-plane vibrations have been recalculated by Gupta et al.,[10] together with their dispersion.

The calculations were carried out according to Wilson's GF matrix method[11] as modified by Higgs[12] for an infinite chain. The methylene groups are treated as mass points (M) since the motions involving these atoms are in the high-frequency region, and they neither disperse much nor are they conformation-sensitive. The molecule of polyglycine I belongs to the factor group C_{2v}. The character table, infrared activity, and number of modes N_i belonging to each symmetry species are given in Table 5.1. Of these, A_1 and B_1 are infrared-active in-plane vibrations and B_2 are infrared-active out-of-plane vibrations. The molecular

Table 5.1. Character Table, Optical Activity, and Symmetry Species of Polyglycine I

C_{2v}	E	C_2	$\sigma_{(zx)}$	$\sigma_{(yz)}$		N_i	
A_1	1	1	1	1	Infrared-active	9	Raman-active
A_2	1	1	−1	−1		4	Raman-active
B_1	1	−1	1	−1	Infrared-active	9	Raman-active
B_2	1	−1	−1	1	Infrared-active	4	Raman-active

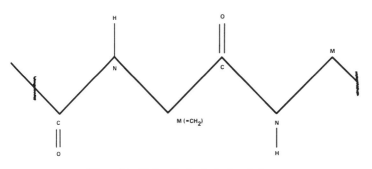

Figure 5.1. Unit cell of polyglycine chain.

model is shown in Figure 5.1. The internal coordinate vectors belonging to a unit, the corresponding local symmetry coordinates, and the matrix of transformation have been calculated by Gupta *et al.*[10]

The torsional coordinates involving motions around the bonds C—N, N—M, and M—C are defined as the sum of the *trans* and *cis* torsions. For example, the coordinate for the torsion around C—N is $\tau_{\text{MCNM}} + \tau_{\text{OCNM}} + \tau_{\text{MCNH}} + \tau_{\text{OCNM}}$. The other two torsions can be similarly defined. The case where only skeletal torsions are defined will be discussed later. Since different values of the force constants should be used for different motions, the values for torsional motions should be taken as average values.

The Wilson *GF* matrix method consists of writing the inverse kinetic energy matrix *G* and the potential energy matrix *F* in internal coordinates *R*. The molecular vibration problem then reduces to a set of equations represented by the matrix equation

$$GFL = L\lambda, \tag{5.1}$$

where the vibration frequencies are given by $\lambda_i = 4\pi^2 \nu_i^2$ and the normal coordinates *Q* by

$$R = LQ. \tag{5.2}$$

In the case of infinite isolated helical polymers, there are an infinite number of internal coordinates which lead to *G* and *F* matrices of infinite order. As pointed out by Higgs, however, in view of the screw symmetry of the polymer a transformation similar to that given by Born and von Kármán can be performed which reduces the infinite problem to finite dimensions.[10]

If δ corresponds to the vibrational phase difference between the corresponding internal coordinates of adjacent chemical repeat units, the vibrational equation becomes

$$G(\delta)F(\delta)L(\delta) = \lambda(\delta)L(\delta), \tag{5.3}$$

where again $\lambda_i(\delta) = 4\pi^2 v_i^2(\delta)$ and the normal coordinates are given as

$$S(\delta) = L(\delta)Q(\delta). \tag{5.4}$$

A calculation of the vibration frequency v_i as a function of δ results in the dispersion relations. It has been shown that $v_i(\delta)$ is symmetric about $\delta = 0, \pi, \ldots$ so that only values for $0 < \delta < \pi$ need be computed.

The above method of calculating the dispersion relations makes use of the screw symmetry of the polymer rather than the translational symmetry of the crystallographic unit cell. Because the model of the polymer used in these calculations is an isolated chain, there exist four zero frequencies, that is, four acoustic modes. These four frequencies correspond to translations along three perpendicular directions and rotation about the long axis of the entire chain.

For the extended chain with the twofold screw axis, the parallel infrared bands are due to the A vibrations with a phase difference of $\delta = 0$ (symmetric with respect to the C_2 axis), whereas the perpendicular bands are due to the B vibrations with a phase difference of $\delta = 180°$ (asymmetric with respect to the C_2 axis).

The Urey-Bradley force field for the molecule of polyglycine I has been derived by Gupta et al.[10] The force constants that give the best fit with the observed infrared frequencies are tabulated in Table 5.2.

Amide bands have been used to characterize conformation changes in polypeptide molecules (Table 5.3). The correlation between chain conformations and amide I and II frequencies was first noticed by Elliot and Ambrose.[13] They observed that the folded conformation exhibits amide I and II bands at about 1650 and 1540 cm^{-1}, whereas the extended conformation exhibits these bands at 1630 and 1520 cm^{-1}. Miyazawa[14] and Miyazawa and Blout[15] have given theoretical interpretations to these correlations and applied them more generally to the random-coil, the α helix, and the antiparallel chain extended conformations. Amide I and II bands, however, are ill defined in some fibrous proteins. The torsional motion around the peptide C—N bond (amide VII), which involves large displacements of the alpha carbon atom, is expected to be most sensitive to the conformational changes of the chain. This band appears as a very strong peak at 210 cm^{-1} in

Table 5.2. Force Constants for Polyglycine I (mdyn/Å) (all angles = 120°)

K(N—H) = 5.45	H(H—N—C) = 0.34	F(H—N—C) = 0.50	K^{τ}_{CN} = 0.315
K(N—C) = 6.23	H(C—N—M) = 0.32	F(C—N—M) = 0.30	K^{τ}_{NM} = 0.120
K(N—M) = 3.58	H(H—N—M) = 0.16	F(H—N—M) = 0.50	K^{τ}_{MO} = 1.30
K(M—C) = 2.91	H(N—M—C) = 0.33	F(N—M—C) = 0.30	K^{ω}_{CO} = 0.280
K(C—O) = 8.90	H(O—C—M) = 0.35	F(O—C—M) = 0.50	K^{ω}_{NH} = 0.120
	H(M—C—N) = 0.32	F(M—C—N) = 0.50	
	H(O—C—N) = 0.35	F(O—C—N) = 1.50	

γ_{NH} = 1.00Å; γ_{NM} = 1.47Å; γ_{MC} = 1.54Å; γ_{CO} = 1.21Å; γ_{CN} = 1.32Å.

101

Table 5.3. Characterization of Amide Bands

Amide I	O—C—N deformation
Amide II	N—H in-plane bending+C—M stretching
Amide III	N—H in-plane bending+C—M stretching+N—C stretching
Amide IV	C—O in-plane bending+C—M stretching+C—N—M bending
Amide V	N—H out-of-plane bending
Amide VI	C—O out-of-plane bending
Amide VII	C—N torsion

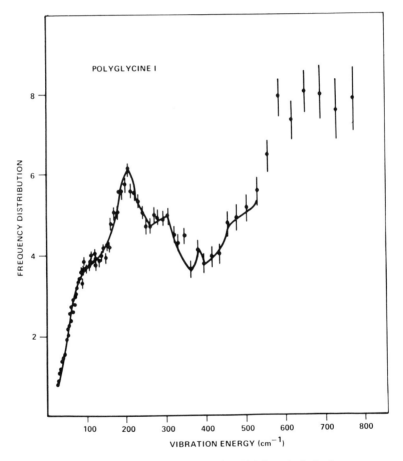

Figure 5.2. Frequency distribution $G(\omega)$ for polyglycine I.

the neutron spectrum of polyglycine I (Figure 5.2) and at 217 cm^{-1} in the infrared absorption.[16] The calculations indicate that this frequency belongs to $\delta = 0$ and, being an out-of-plane mode, it should be infrared-inactive. The corresponding frequency belonging to $\delta = \pi$ is calculated at 410 cm^{-1} and should be infrared-active. Indeed a sharp absorption band at 410 cm^{-1} is observed in infrared. There are two explanations for the appearance of the band at 215 cm^{-1}. Either it belongs to a $\delta = 0$ torsional mode and arises because of the nonplanarity of the polyglycine I molecule, or it belongs to the in-plane mode calculated at 221 cm^{-1}.[9, 12] Nonplanarity is the more probable reason since this band appears very strong in the neutron spectra, where intensities are weighted by the amplitude of motions of the hydrogen atoms. This torsional motion, which according to the calculated potential energy distribution is made up of τ_{MC} (61%) + τ_{NM} (15%) + τ_{CN} (10%) + τ_{NH} (14%), would necessarily cause large hydrogen motions. The principal contributions to the in-plane mode calculated at 221 cm^{-1} are def$_{MCN}$ (38%) + def$_{NMC}$ (11%) + def$_{CNM}$ (14%) + def$_{MCO}$ (13%), and these do not involve large displacements of hydrogen atoms. In polyglycine II, where the chain exists as a threefold helix and the hydrogens of the methane group are bonded to the nearest neighbor via C—H---O—C bonds, the displacement amplitude of the hydrogens in torsional motion is much reduced; hence the corresponding band in neutron spectra is relatively weak. This band appears at 350 cm^{-1} (Figure 5.3) in good agreement with infrared results (365 cm^{-1}).[16] In other words, the reduced intensity of this torsional mode in the spectra independently supports the idea put forth recently by Ramachandran[5] that, as in the structure of collagen, hydrogen bonds of the type C—H---O can be formed in polyglycine II. The amplitude of motion of the hydrogen in polyglycine II must therefore be reduced because of CH---O bonding and not because of change in conformation. This is confirmed by the fact that the torsional band in question is again very intense in the neutron spectra of polyalanine.[12] The dispersion curves of a polyglycine I chain are shown in Figure 5.4; they result from the calculations outlined at the beginning of this chapter.

It is interesting to note that the dispersion curves for the low-frequency optical and acoustic modes show flat regions in the frequency range 100–140 cm^{-1}. This implies a high density of states at these frequencies and is reflected in the shoulder in the inelastic neutron spectra of polyglycine I. Of course, in this region there are also lattice frequencies corresponding, for example, to N—H---O stretching.

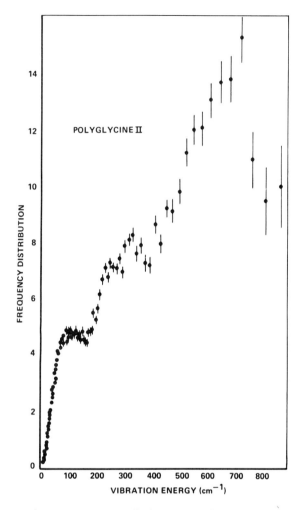

Figure 5.3. Frequency distribution $G(\omega)$ for polyglycine II.

Because of the selection rules, only modes corresponding to $\delta = 0$ are observed in infrared absorption; inelastic neutron scattering data, on the other hand, are not governed by any selection rules and reflect the density of states. The observations of low-frequency modes are important in determining such thermodynamic properties as specific heat, which is largely dependent on the distribution of these modes.

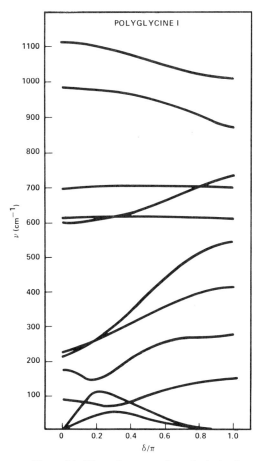

Figure 5.4. Dispersion curves for polyglycine I.

5.2 Calculation of Thermodynamic Quantities

A knowledge of the frequency distributions for a sample permits the calculation of such thermodynamic quantities as free energy, enthalpy, entropy, and specific heat, the sample being treated as an assembly of harmonic oscillators. Recently[17] use was made of the frequency distribution for polyglutamic acid (helix) and its sodium salt (random coil) to calculate the energy of formation of the intramolecular hydrogen bond that stabilizes the helix. Using Equation 3.57 together with the frequency distributions $G(\omega)$ obtained from neutron scattering for polyglycine I and polyglycine II and correcting for the difference in the zero-point energies of the two forms, a difference of 0.5 kcal/residue was found, the energy being higher for polyglycine I than for polyglycine II. This value should be the sum of the conformational energy differences and the lattice energy differences between the two. The lattice energy is made up of the N—H---O hydrogen bonds, an extra C—H---O hydrogen bond present only in polyglycine II, and the nonbonded interactions. According to Scott and Scheraga's conformational analysis of macromolecules,[18] it would seem that the difference between the conformational energy of polyglycine I (antiparallel-chain pleated sheet, $\phi = 38°$ and $\psi = 325°$) and that of polyglycine II ($\phi = 100°$, $\psi = 330°$) is not very significant. The N—H---O type hydrogen bonds in the two forms are also of about the same strength. Assuming that only one-third of the possible C—H---O bonds exist in polyglycine II, as has been theoretically suggested by Ramachandran[8] (experimentally this fraction may vary from sample to sample), the calculations indicate that the energy of formation of the C—H---O bond in polyglycine II is less than 1.5 cal/residue. It should be pointed out that these calculations are based on a limited frequency distribution (up to 900 cm^{-1}) as obtained by neutron scattering; moreover, the frequency distribution for both forms should really be weighted by the polarization vector for each mode. The difference in the frequency distribution for higher modes is not expected to be very significant, but the polarization vectors could make a sizable difference.

5.3   Torsions Around the C—N Bond

Finally, calculations of the dispersion curves for polyglycine I for the two definitions of torsion—that is, as a sum of all *cis* and *trans* torsions and as only skeletal torsions—indicate that although the dispersion of torsional modes around the peptide C—N bond is not very different in the two cases, the potential energy distribution (PED) into various modes is very much altered. In both cases the torsion around the C—N bond mixes with the torsion around N—M and M—C bonds and also with the C—O and N—H out-of-plane waggings. For the phase difference $\delta = 0$, the PED in the C—N torsional mode is only 10 percent in the first case and 33 percent in the second; with increasing phase the PED for $\delta = \pi$ becomes 25 percent in the first case and 40 percent in the second.

This torsional motion around the peptide C—N bond (also called amide VII) involves a somewhat larger displacement of the carbon atoms; accordingly, strong vibrational interactions take place between similar vibrations of adjacent groups. Moreover, for nonplanar conformations of polypeptides vibrational interactions between the C—N torsional modes and all the valence and angle-bending modes depend sensitively on helical conformations. The amide VII, because of its low frequency, has not been easily accessible to infrared measurement; it has been observed in only a few cases. In *N*-methylacetamide, for example, it was observed at 205 cm^{-1}, in good agreement with the neutron data;[20] this would correspond (assuming a threefold cosine-type barrier) to a barrier height of about 14 kcal/mole. If the double-bond barrier is 40 kcal/mole, the partial double-bond character of the C—N bond is estimated at about 0.3 to 0.4, in agreement with the values estimated from the C—N and C—O bond lengths.[21] Amide VII frequencies have been observed by neutron scattering in a large number of substances of biological interest.[20] For a series of glycine compounds, it has been possible to show[10] the existence of this amide VII frequency and its sensitive dependence upon the conformation of the molecule. Glycine

$$(^+NH_3—CH_2—\overset{\overset{\displaystyle O}{\|}}{C}—O^-)$$

does not contain the peptide group, but glycyglycine, triglycylglycine, and polyglycine I (transplanar) contain an increasing number of such

Table 5.4. Vibration Frequencies (cm^{-1}) in Glycylglycine

Neutron	Infrared	Assignment
100		lattice modes
160		lattice modes
200		C—N torsion (Amide VII)
250		C—C torsion
300	317	C—C—N$^+$ bending
380		HN—C—C bending
435		C—N—C bending
500	531	COO$^-$ rocking and NH$_3^+$ torsion
600	591	Amide IV and Amide VI
710	730	Amide V
900		CH$_2$ rocking

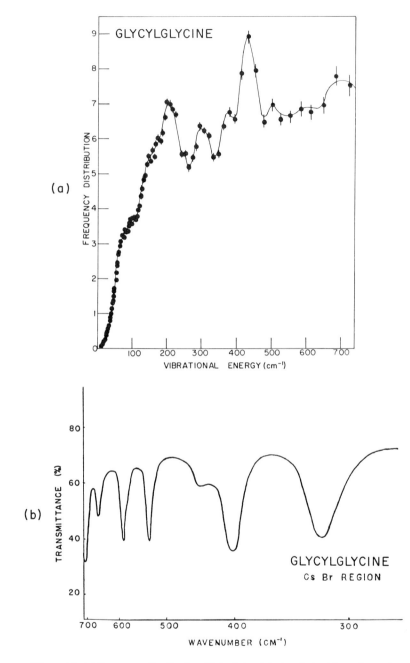

Figure 5.5. (a) Frequency distribution $G(\omega)$ of glycylglycine derived from the neutron data. (b) Infrared spectrum of glycylglycine in the CsBr region.

108

peptide groups. A sharp peak at 200 cm^{-1} is observed in the neutron spectra of glycylglycine and of all higher polymers of glycine but not in the neutron spectrum of glycine. Moreover, this peak increases in intensity with the number of peptide groups in the molecule. It is therefore assigned the amide VII frequency.

The frequency distribution $G(\omega)$ of glycylglycine is shown in Figure 5.5, together with the infrared spectrum above 300 cm^{-1}. Table 5.4 provides possible assignments of the low-frequency peaks in glycyl-glycine.

REFERENCES

1. C. W. Bunn and E. V. Garner, *Proc. Roy. Soc. (London)* **A189**, 39 (1947).
2. F. H. C. Crick and A. Rich, *Nature* **176**, 780 (1955).
3. F. J. Padden and H. D. Keith, *J. Appl. Physics* **36**, 2987 (1965).
4. S. Krimm, *Nature* **212**, 1482 (1966).
5. C. N. Ramachandran, V. Sasisekharan, and G. Ramakrishnan, *Biochim. Biophys. Acta* **112**, 168 (1966).
6. G. N. Ramachandran and V. Sasisekharan, *Biochim. Biophys. Acta* **109**, 314 (1965).
7. S. Krimm, K. Kuroiwa, and T. Rebane, Paper presented at the conference on Conformation of Biopolymers held at Madras, India, 1967.
8. G. N. Ramachandran, G. Ramakrishnan, and C. M. Venkatachalam, Paper presented at the conference on Conformation of Biopolymers held at Madras, India, 1967.
9. K. Fukushima, Y. Ideguchi, and T. Miyazawa, *Bull. Chem. Soc. Japan* **36**, 1301 (1963).
10. V. D. Gupta, S. Trevino, and H. Boutin, *J. Chem. Phys.* **48**, 3008 (1968).
11. E. B. Wilson, *J. Chem. Phys.* **7**, 1047 (1939).
12. P. W. Higgs, *Proc. Roy. Soc. (London)* **A220**, 472 (1953).
13. A. Elliot and E. J. Ambrose, *Nature* **165**, 921 (1950).
14. T. Miyazawa, *J. Chem. Phys.* **32**, 1647 (1960).
15. T. Miyazawa and E. R. Blout, *J. Am. Chem. Soc.* **83**, 712 (1961).
16. T. Miyazawa, *Bull. Chem. Soc. Japan* **34**, 691 (1961).
17. H. Boutin and W. L. Whittemore, *J. Chem. Phys.* **44**, 3127 (1966).
18. R. A. Scott and H. A. Scheraga, *J. Chem. Phys.* **45**, 2091 (1966).
19. G. Nemethy, G. N. Ramachandran, and H. A. Scheraga, *J. Biol. Chem.* **241**, 1004 (1966).
20. H. Boutin, Technical Rept. 3367, Picatinny Arsenal, Dover, New Jersey, 1966.
21. T. Miyazawa, in *Polyamino Acids, Polypeptides, and Proteins,* edited by Mark A. Stahmann, The University of Wisconsin Press, Madison, Wisconsin, 1962, p. 201.

6. Hydrogen-Bond Vibrations

6.1 KHF_2, KH_2F_3, HF, and HCl

The experiments discussed in this chapter illustrate that there are. in fact, "selection rules" in the neutron-scattering process, although they were not emphasized in the theoretical considerations of Chapters 2 and 3. Figure 6.1 shows the neutron spectrum of a polycrystalline sample of KHF_2.[1] Two well-resolved peaks at 1176 and 600 cm^{-1} are assigned to the fundamental vibration v_1 (bending) and v_2 (stretching) of the HF_2 ion. This ion has a well-established, symmetric F—H—F bond (F—F distance 2.29Å). The bending vibration of the hydrogen bond is expected to have large amplitude, and this gives rise to a sharp peak in the neutron spectrum. The hydrogen-bond stretch, because of the symmetry of the ion, does not involve any motion of the hydrogen atom and therefore appears as a weak peak in the neutron spectrum. This may constitute a type of neutron selection rule of possible value in establishing the symmetry of the hydrogen bond. In KH_2F_3, for example, where the ion is known to possess a strong hydrogen bond (F—F distance 2.39Å), its symmetry has never been established. The neutron spectrum[1] (Figure 6.2) is quite similar to that of KHF_2 with prominent peaks at 896 and 360 cm^{-1}. The peak at 896 cm^{-1} is assigned to a hydrogen-bond bending motion; it is at lower frequency than in KHF_2 because of an increased F—F distance. The peak at 360 cm^{-1} is assigned to hydrogen-bond stretching. The sharpness of the latter suggests that in the case of KH_2F_3 the H bond is not sym-

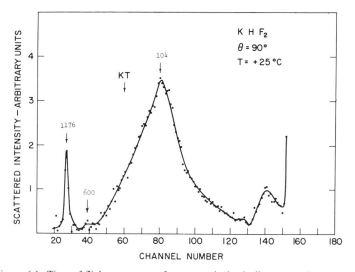

Figure 6.1. Time-of-flight spectrum of neutrons inelastically scattered at 90° angle from a polycrystalline sample of KHF_2. The incident neutrons scattered elastically correspond to the line at channel 152. The spectrum observed between channel 20 and 130 corresponds to the inelastic scattering and the number above each peak is the energy transfer in cm^{-1}. The ordinates are proportional to the differential cross section.

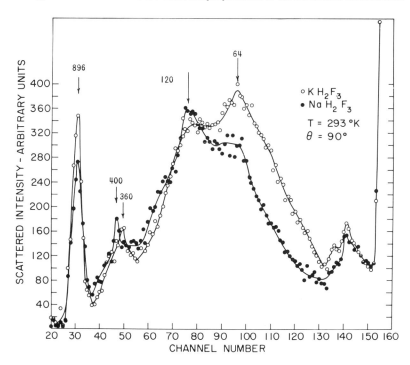

Figure 6.2. Time-of-flight spectrum of neutron inelastically scattered by polycrystalline samples of KH_2F_3 and NaH_2F_3. The energy transfers in cm^{-1} are indicated above each peak. The ordinates are proportional to the differential scattering cross section.

111

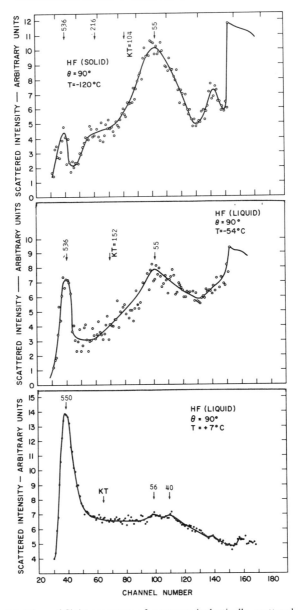

Figure 6.3. Time-of-flight spectrum of neutrons inelastically scattered by a poly-crystalline sample of HF and for HF liquid at two temperatures. The energy transfers in cm^{-1} are indicated above each peak. The ordinates are proportional to the differential scattering cross section.

metric but rather that the hydrogen is associated with one or the other of the fluorine atoms. A similar case is that of HF, whose neutron spectra in the solid and liquid phases[1, 2] are shown in Figure 6.3. The broad but pronounced shoulder centered around 250 cm^{-1} in the neutron spectrum of the solid phase is assigned to the translational vibrations. These are modes in which adjacent hydrogen-bonded molecules "beat" against each other, the polarization being either parallel or perpendicular to the chain axis. In the infrared spectrum of solid HF (Figure 6.4)[3], two bands at 202 and 366 cm^{-1} which do not shift upon deuteration are assigned to these modes. They are not resolved in the neutron spectrum, but the intensity of the shoulder at 250 cm^{-1} is much greater than the corresponding peak (600 cm^{-1}) in KHF₂ because the hydrogen bond in HF has been shown by neutron

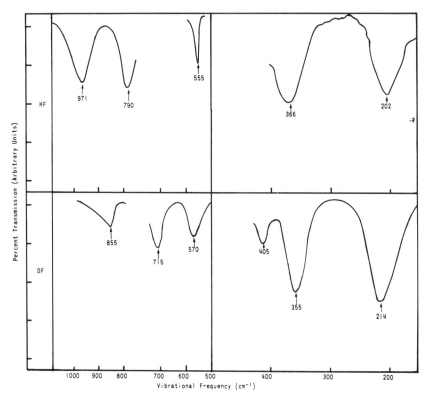

Figure 6.4. Infrared and far infrared spectra of HF and DF.

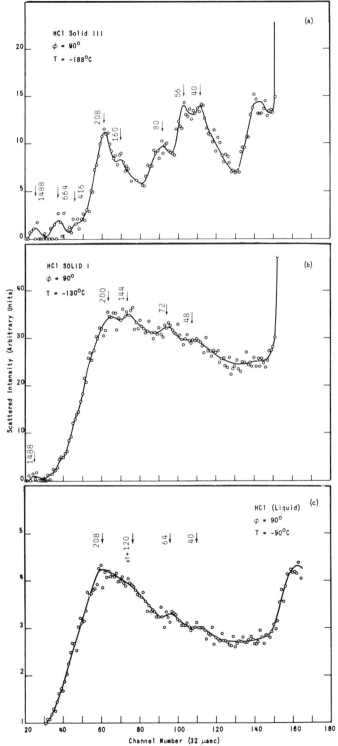

Figure 6.5. Time-of-flight spectra of neutrons scattered inelastically by a poly-crystalline sample of HCl at (a) 85°K and (b) 143°K and (c) by a liquid sample of HCl at 183°K. The energy transfers in cm^{-1} are indicated above each peak. The ordinates are proportional to the scattering cross sections.

diffraction to be quite asymmetric (F—F distance = 2.49Å). The strong peak at 536 cm^{-1} in the neutron spectra of the solid and liquid phases of HF is assigned to an out-of-plane libration of the molecule in the zigzag planar chain. This motion is also observed in the infrared spectrum of solid HF at 555 cm^{-1} and shifts to 405 cm^{-1} upon deuteration, thereby justifying its assignment. The persistence of this peak without appreciable broadening in the liquid phase up to the boiling point[1] is taken as an indication of the existence of large polymeric units in liquid HF. A drastic broadening of this peak is observed in the neutron spectrum of HCl,[1] for example, where polymeric units are rapidly broken with increasing temperature. In HCl, the neutron spectra at three temperatures[1] (Figure 6.5) display a strong peak around

Table 6.1. Comparison of the Frequencies of the Lines Observed by Far Infrared and Neutron Inelastic Scattering for Solid HCl and Solid HBr (in cm^{-1})

HCl		
Anderson et al.[a] Phase III 77°K	Hornig and Osberg[b] Phase III 77°K	This Work Phase III 100°K
		40
		56
86		80
109		
		152
217		208
296		(280)
	496	420
	650	660

HBr				
Anderson et al.[a]		Hornig and Osberg[b]	This Work	
Phase II	Phase III	Phase III	Phase I	Phase II
				24
57	57		56	56
71	71			(78)
			128	128
	200			
218			184	184
	269		272	
		400		
		575		

[a] A. Anderson, H. A. Gebbie, and S. H. Walmsley, *Mol. Phys.* **7**, 401 (1964).
[b] D. F. Hornig, W. E. Osberg, *J. Chem. Phys.* **23**, 662 (1955).

200 cm^{-1} which is assigned to out-of-plane libration and is in agreement with far-infrared data.[4] This peak corresponds to the HF peak observed at 536 cm^{-1}, the lower frequency being due to weaker hydrogen bonds. This librational band broadens considerably with increasing temperature, indicating a rapid disruption of the zigzag chains. A comparison of neutron and far-infrared frequencies for HCl and HBr is shown in Table 6.1.

6.2   HCrO$_2$

In HCrO$_2$, two prominent peaks are observed in the neutron spectrum[5] at 475 and 225 cm^{-1}. Because of their intensity they must involve large amplitude motions of the H bond. The peak at 475 cm^{-1} remains intense and does not shift upon deuteration (neither do the corresponding infrared frequencies at 520 cm^{-1} in HCrO$_2$ and 505 cm^{-1} in DCrO$_2$), so it is probably translational in nature. If the peak at 225 cm^{-1} is assigned to an intermolecular hydrogen-bond-stretching vibration, its respective intensity in HCrO$_2$ and DCrO$_2$ would suggest that both are asymmetric. However, this conclusion, derived from examination of the neutron data, is contrary to the infrared results which suggest that DCrO$_2$ has a noncentrosymmetric structure and asymmetric hydrogen bonds (all optical modes are active) while HCrO$_2$ has effectively symmetric O—H—O bonds.

6.3   KH$_2$PO$_4$ (Potassium Dihydrogen Phosphate)

An extensive study of KH$_2$PO$_4$ has been performed by Imry *et al.*[6] It will be presented here in some detail since it illustrates the importance of using neutron data in correlation with infrared and Raman measurements. In KH$_2$PO$_4$-type crystals it is known that the motion of the protons is responsible for the ferroelectric transition. The involvement of the protons is seen, for example, from the large shift in the Curie point upon deuteration.[7] Moreover, the hydrogen bond, being the weakest bond in the crystal, is probably also responsible for the low melting point of these crystals (for example, 253°C for KH$_2$PO$_4$ compared to 1340°C for K$_3$PO$_4$). The neutron diffraction data of Pease and Bacon[8] on KH$_2$PO$_4$ show a symmetric elongated proton distribution along the axis of the hydrogen bond above the Curie temperature and a more concentrated distribution below it. There are several ways of explaining their results above the Curie point: the proton may be located asymmetrically near one of the oxygens but be statistically disordered in the crystal sites, or the proton may actually be near both oxygens as a result of one of several possible anharmonic motions. Further information about the motions of the protons can be gained by studying the spectral consequences of these hypotheses, using the various techniques of spectroscopy (Raman, infrared, neutron scattering). Indeed, a comparison of measurements of the same phenomenon obtained by the different techniques[9] can provide much additional insight. In the case of KH_2PO_4 crystals, a comparison of the different spectra indicates the existence of very broad low-energy hydrogen modes that appear prominently only in neutron-inelastic-scattering data. The existence of these low-proton modes supports the picture that in the nonferroelectric phase there exists on the average a slightly asymmetric double-minimum potential well where the protons tunnel quantum-mechanically. This picture suggests that the cooperative effects of the protons can be taken care of by the asymmetry parameter in the proton double-minimum potential. The treatment of such a system shows that in the temperature region between the ferroelectric Curie point and the melting point it is possible to get negative thermal expansion in the hydrogen-bond direction. The two phase-transition points bear an internal relation to each other, determined by the H bonds. Experimental evidence for this is found in the indication of a rise in the dielectric constant as the melting temperature is approached. In the light of these considerations, Wiener and Pelah measured the infrared spectrum just above and just below the Curie point and found significant changes,[10] especially in the 400 cm^{-1} region. Measurements of the deuterated analog KD_2PO_4 support this picture and make the assignment of the 2400–2800 cm^{-1} doublet as hydrogen vibrations[11,12] questionable. The spectra of K_2HPO_4 obtained by neutron inelastic scattering and infrared absorption (Figure 6.6) are very similar in the 250–700 cm^{-1} region. One observes two rather sharp peaks at 540 and 420 cm^{-1}, which were assigned[6] to the characteristic PO_4 modes v_3 and v_2, respectively. In K_3PO_4 the same peaks were found in the infrared spectrum,[6] but they are absent in neutron-inelastic-scattering data.[6] Since the difference between K_3PO_4 and K_2HPO_4 is that the hydrogen is replaced by potassium, it is assumed that the peaks in the neutron-inelastic-scattering spectra of K_2HPO_4 appear predominantly because of the hydrogen that is bonded to one of the oxygens of the PO_4 group and participates in its characteristic vibrations. In the infrared spectra of KH_2PO_4 two peaks are observed:

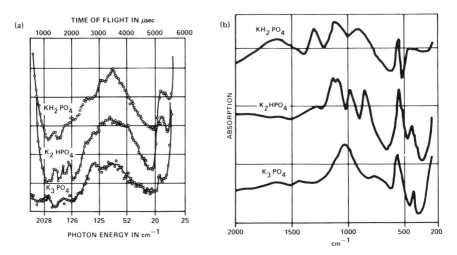

Figure 6.6. (a) Neutron time-of-flight spectra of K_3PO_4, K_2HPO_4, and KH_2PO_4 at room temperature. (b) Infrared spectra of K_3PO_4, K_2HPO_4, and KH_2PO_4 at room temperature for the wave-number region 250 to 2000 cm^{-1}.

a sharp one at 530 cm^{-1} and a broad one from 450 to 250 cm^{-1}. On the other hand, in neutron-inelastic-scattering spectra one observes just one broad peak extending from 600 to 250 cm^{-1}. Since the resolution is good enough to separate peaks in this region, as seen in the case of K_2HPO_4, these results indicate that the protons are also involved in low-energy modes that are not PO_4-characteristic; these modes extend from 600 to 250 cm^{-1} and appear predominantly in neutron inelastic scattering but not in infrared absorption. Extremely broad low-energy proton transitions of this type can be accounted for[9] by assuming the existence, on the average, of a slightly asymmetric double-minimum potential well along the O—H—O bond, where the proton tunnels quantum-mechanically. The vibrational ground level of the proton is split into two close levels with a separation of about 400 cm^{-1}. It seems that the small overlap in the wave functions weakens transitions between the split levels in infrared and Raman scattering but not in neutron inelastic scattering.

Using the Raman scattering technique, Narayanan[13] and Chapelle[14] observed the PO_4-characteristic modes in the region 250 to 700 cm^{-1} but did not find a pronounced broad protonic peak; however, it is possible that the weak diffused peak seen by Chapelle[14] below 500 cm^{-1} is this protonic transition. A comparison of the spectra of KH$_2$PO$_4$ and KD$_2$PO$_4$ in the region 450 to 250 cm^{-1} does show a difference in shape, KD$_2$PO$_4$ having two peaks at 380 and 450 cm^{-1} while KH$_2$PO$_4$ has just one broad peak covering both. This suggests that part of the KH$_2$PO$_4$ absorption in this region is due to the broad protonic mode centered between the PO$_4$ modes. On the other hand, in KD$_2$PO$_4$, as discussed later, the deuteronic transition would be highly forbidden and therefore hardly seen in the infrared. Other measurements made by Barker and Tinkham[15] on the far-infrared reflections from KH$_2$PO$_4$ have shown a dependence of the 400-cm^{-1} band on direction, which may be explained as arising from the preferred directions in the proton movements in the crystal.

The tunneling motions of the hydrogen may interact with some of the PO$_4$-characteristic modes and broaden them, especially when these modes have a component of motion in the H-bond direction and energies comparable with the tunneling mode. In the ferroelectric phase the hydrogens order themselves and tunneling motions stop. Therefore, one expects a distinct sharpening of the low PO$_4$ modes along with the disappearance of the hydrogen tunneling mode below the Curie point. The infrared spectra show that this is indeed the case for the 400-cm^{-1} region. Just below the transition temperature in KH$_2$PO$_4$ the broad 400-cm^{-1} peak turns into two sharp peaks, while in KD$_2$PO$_4$ the same two peaks sharpen markedly. Changes during phase transition in other regions of the spectra are less pronounced and are due to the lowering of crystal symmetry. The changes in spectra with temperature in both phases are otherwise very small.

The experimental results of Imry *et al.*[6] point to the existence, above the Curie point, of a broad protonic level at about 400–500 cm^{-1}. It is tempting to ascribe this level to the splitting of the ground state in a double-minimum potential well. Some asymmetry of the potential well is assumed in order to account for the infrared selection rules. The asymmetry has a relatively small effect on the energy levels.

The low protonic level is explained on the basis of a proton moving in an isolated potential well. This picture can be justified as an approximation in the following way. It is well known that short-range interactions do exist between the motions of the protons. These interactions tend to produce short-range ordering of the protons in their sites, that is, exactly two protons are located near any PO$_4$ group.[16] Thus, the double-minimum potential well of any proton tends to be highly asymmetric. It is possible to picture each proton as being in a potential field with an average asymmetry arising from the correlation with other protons and changing with time. The NMR experiments of Schmidt and Uheling[17] show that the asymmetry changes its sign with a frequency of 10^{11}–10^{12} sec^{-1}. This frequency is much lower than the typical frequency of the protonic motion in the instantaneous well, so that the problem can be treated adiabatically for the motion of an isolated proton. The observed broadness of the low protonic level is partly accounted for by this picture since the levels change with the instantaneous potential. The interactions with other vibrations (for example, those of the PO_4 group) also contribute to the width of the level.

An interesting qualitative result of this picture is that the average asymmetry increases with O—O distance. When the O—O distance increases, the O—H distance decreases[18] and the short-range interactions between the protons increase. At the same time, the tunneling probability of the proton decreases (because of the larger potential barrier) and the potential becomes less symmetric. The same result—namely, that asymmetry is a monatomically increasing function of O—O distance—was obtained in a semiempirical treatment of the hydrogen bond by Reid.[19] He also predicts a frequency of 240 cm^{-1} for the O—H—O stretching vibration.

Below the Curie point it is expected that the static polarization will cause the protonic potential to be extremely asymmetric. Each proton will remain at a specific site and both the tunneling and the slow correlated motion of the minima will disappear. This is evidenced by both NMR[17] and IR[6] results. The v_2 vibrations of the PO_4,[10] which are seen quite sharply below the Curie point, are broadened by the hydrogen tunneling motion above the Curie point and appear as a broad peak. The broad tunneling level which is only weakly allowed in infrared might also contribute to the absorption intensity above the Curie point.

6.4 Other Ferroelectrics

Low-frequency motions in several ferroelectric salts have also been studied by Rush and Taylor.[20] Energy-gain spectra for the inelastic scattering of cold neutrons by ferroelectric $K_4Fe(CN)_6 \cdot 3H_2O$, NH_4HSO_4, $(NH_4)_2SO_4$, and $(NH_4)_2BeF_4$ have been measured above and below their Curie points. For $K_4Fe(CN)_6 \cdot 3H_2O$ ($T_c = 249°K$), scattered spectra were obtained at 121, 175, and 296°K. The results were quite similar in both the high-temperature and the ferroelectric phases. A broad band centered at an energy gain of 435 ± 40 cm^{-1} (54 ± 5 meV) is observed at all three temperatures and assigned to the librations of the water molecules. A second broad peak around 165 cm^{-1} is possibly due to the translational motions of the water molecules in the lattice.

The room-temperature spectra for NH_4HSO_4 (T_c at 154 and 270°K) and for $(NH_4)_2SO_4$ ($T_c = 224$°K) show broad bands peaked at about 260 cm^{-1} and 300 cm^{-1}, respectively, which are primarily due to the torsional motions of the ammonium ions. The NH_4HSO_4 spectrum at 177°K shows an indication of splitting, with little shift in the center of the band. At 125°K the spectrum is clearly split, with peaks at 290 ± 25 and 190 ± 16 cm^{-1}. The $(NH_4)_2SO_4$ results at 172°K also exhibit a split band, with peaks centered at 335 ± 25 and 200 ± 16 cm^{-1}. In both cases the higher-energy peaks are assigned to torsional vibrations. The $(NH_4)_2BeF_4$ spectra also show broad "torsional" bands with several indicated maxima in the 175–290 cm^{-1} range. No great differences were observed in the spectra above and below the ferroelectric transitions for any of the ammonium salts.

The results for all the compounds studied indicate that the ferroelectric transitions are not related to any large change in the average rotational freedom of the ammonium ions and water molecules. Moreover, the measurements showed that the average barriers to reorientation are relatively small ($V_0 \gtrsim 4$ kcal/mole) in every case. These conclusions do not preclude changes in the ordering of the protons at the transitions.

REFERENCES

1. H. Boutin, G. J. Safford, and V. Brajovic, *J. Chem. Phys.* **39**, 3135 (1963) and references therein.
2. H. Boutin and G. J. Safford, in *Inelastic Scattering of Neutrons in Solids and Liquids* (International Atomic Energy Agency, Vienna, 1965), Vol. 2, p. 393.
3. M. L. N. Sastri and D. F. Hornig, *J. Chem. Phys.* **39**, 3497 (1963).
4. A. Anderson, S. H. Walmsley, and H. A. Gebbie, *Phil. Mag. (Ser. 8)* **7**, 1243 (1952).
5. J. J. Rush and J. R. Ferraro, *J. Chem. Phys.* **44**, 2496 (1966).
6. Y. Imry, I. Pelah, and E. Wiener, *J. Chem. Phys.* **43**, 2332 (1965).
7. F. Jona and G. Shirane, *Ferroelectric Crystals* (Pergamon Press, London, 1962).

8. R. S. Pease and G. E. Bacon, *Proc. Roy. Soc. (London)* **A220**, 397 (1953); **A236**, 359 (1955).

9. I. Pelah, E. Wiener, and Y. Imry, in *Inelastic Scattering of Neutrons from Solids and Liquids* (International Atomic Energy Agency, Vienna, 1965), Vol. 2, p. 325.

10. E. Wiener and I. Pelah, *Phys. Letters* **13**, 206 (1964).

11. R. Blinc and D. Hadzi, *Mol. Phys.* **1**, 391 (1958).

12. A. N. Lazarov and A. S. Zaitseva, *Soviet Phys—Solid State* **2**, 2688 (1960).

13. P. S. Narayanan, *Proc. Indian Acad. Sci.* **A233**, 240 (1951).

14. J. Chapelle, *J. Chim. Phys.* **49**, 30 (1949).

15. A. S. Barker and M. Tinkham, *J. Chem. Phys.* **38**, 2257 (1963).

16. J. C. Slater, *J. Chem. Phys.* **9**, 16 (1941); P. G. de Gennes, *Solid State Commun.* **1**, 132 (1963).

17. V. H. Schmidt and E. A. Uheling, *Phys. Rev.* **126**, 2, 447 (1962).

18. G. C. Pimentel and A. L. McClellan, *The Hydrogen Bond* (W. H. Freeman and Company, San Francisco, California, 1960), p. 259.

19. C. Reid, *J. Chem. Phys.* **30**, 182 (1959).

20. J. J. Rush and T. I. Taylor, in *Inelastic Scattering of Neutrons in Solids and Liquids* (International Atomic Energy Agency, Vienna, 1965), Vol. 2, p. 333.

7. Alkaline Earth and Alkali Hydroxides

7.1 Interpretation of Spectroscopic Data

Comparisons between spectra obtained with neutron and optical spectroscopy clearly illustrate the potential value of this method for clarifying lattice-interaction problems.

The interpretation of infrared absorption spectra for alkali and alkaline earth hydroxides has led to considerable controversy in the literature.[1-4] The discussion centers around the near-infrared side bands of the fundamental OH-stretching vibration. From temperature-dependent measurements it has been concluded that the observed spectra are due to combination bands.[5] Hexter and Dows[6] have suggested that these bands are caused by the interaction between libration and vibration in the OH ion, in a way similar to the well-known rotation-vibration interaction in polyatomic gases. Hexter emphasized the "localized" nature of the OH motion, assuming essentially that all OH ions vibrate without phase correlations between one another.

Contrary to this view, Wickersheim[7] and Buchanan *et al.*[2] have explained the hydroxide absorption spectra as being due to interaction between internal and lattice modes. In this interpretation, the phase-correlated motions of the ions are the important ones. Their arguments are supported by far-infrared and Raman studies[8-11] showing directly most of the peaks observed previously only in combination with the stretching frequency. Further support of the lattice approach can be

123

found in inelastic-neutron-scattering measurements by Safford *et al.* In their first note,[12] these authors discuss their spectra on the basis of

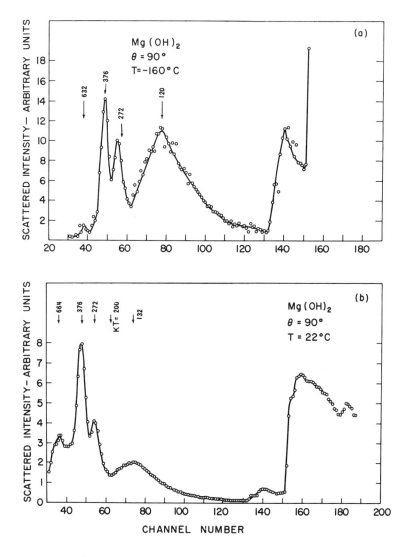

Figure 7.1. Time-of-flight spectrum of neutrons inelastically scattered by a poly-crystalline sample of $Mg(OH)_2$ at (a) 22°C and (b) -160°C. The energy transfers in cm^{-1} are indicated above each peak. The ordinates are proportional to the differential scattering cross section.

Hexter's uncoupled librator theory. However, in a later publication,[13] they state that the theory is ultimately not suitable for explaining their experimental results. The neutron spectra of $Mg(OH)_2$ at two temperatures are shown in Figures 7.1(a) and (b). More recently, neutron data have also been obtained for LiOH by Safford and LoSacco[14] and by Pelah *et al.*[15] The neutron data for deuterated samples have also been presented (Figure 7.2).

It should be initially pointed out that there is relatively good agreement between optical and neutron data, especially if one considers the rather limited experimental resolution in neutron and near-infrared fine-structure measurements. It is interesting that this agreement applies not only to the position of the peaks but also to the shapes of the neutron spectra, which resemble the near-infrared fine-structure curves. This similarity suggests the possibility that in both cases the spectra are caused by the same phenomenon.

In the scattering experiments, neutrons gain energy from interaction with the crystal. The main sources of stored energy in the low-frequency range are lattice vibrations, which at low frequencies give rise to the gradually increasing intensity distribution characteristic of neutron-scattering measurements on incoherent scatterers. The same behavior is observed in this case. It is assumed, therefore, that the neutron-scattering curves also reflect the lattice frequency spectrum of each substance. Under this assumption the good agreement in position and shape between the optical-combination bands and the neutron spectra is considered a good argument for a lattice interpretation of the infrared side bands. Moreover, in the optical case there is an interaction with lattice vibrations of all momentum values, as is the case for incoherently scattered neutrons. This fits well into the framework of the general theory for coupling between internal and lattice modes.[16-18]

Because of their shape and position, the low-energy peaks in the neutron spectra are probably caused by acoustic (translation) vibrations [in LiOH, LiOD, and $Mg(OH)_2$ at about 125 cm^{-1}; in $Ca(OH)_2$ at about 90 cm^{-1}]. The same should be true for the corresponding optical peaks. This point of view is further supported by the experimental fact that these peaks have been observed in optical measurements only as combination bands but not as fundamentals.[2, 11] Some of the observed high-energy peaks must be due to the usual optical translation vibrations which appear in crystals with more than one atom per unit cell. In all three hydroxides these modes appear above 200 cm^{-1}.

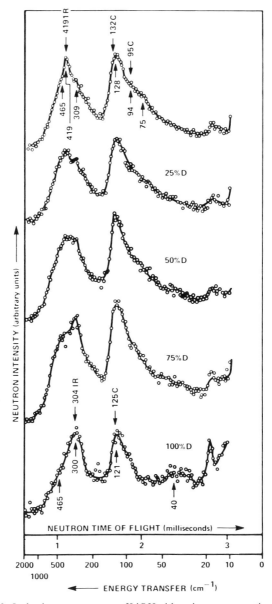

Figure 7.2. Inelastic neutron spectra of LiOH with various concentrations of deuterium. The arrows marked C are obtained by dividing the difference of corresponding near-infrared combination bands by 2. The arrows marked with IR refer to infrared measurements.

7.2 Comparison of Spectroscopic Data for Various Hydroxides

In comparing the data for LiOH, $Mg(OH)_2$, and $Ca(OH)_2$, two features should be noted:

1. The number of peaks and their relative intensities are similar in $Mg(OH)_2$, $Ca(OH)_2$, and LiOH.

2. There is no over-all coincidence of lines in the three compounds; distinct shifts in frequency are observed, presumably due to the influence of the different cations. However, these shifts are not large. Some near coincidences do exist, within experimental error, between lines (at 625 and 633 cm^{-1}) and between acoustic modes (at 138 and 135 cm^{-1}) in LiOH and $Mg(OH)_2$, respectively.

The relatively good agreement between the infrared lines and the neutron peaks observed at 625, 468, and 332 cm^{-1} suggests that these modes have narrow frequency dispersions which are outside the resolution of neutron-scattering experiments. It should be noted that all the phases of a mode are observed in neutron experiments, whereas only certain optically active phases are observed in infrared measurements. Infrared lines at 3823, 3906, and 3762 cm^{-2} have been associated with modes at 152, 135, and 95 cm^{-1} combining with the OH^- fundamental. These lines have not been seen in direct infrared measurements. The data for LiOH show that the side bands with separations of 152 and 135 cm^{-1} both appear to be associated with the acoustic mode having its high-frequency limit near 138 cm^{-1}, while that of 95 cm^{-1} may be associated with the observed shoulder at 85 cm^{-1}.

The individual peak intensities vary according to the temperature dependence of the one-phonon neutron incoherent-scattering cross section.[19] However, the observed peaks at a given temperature do not show the systematic intensity variation that would be expected for consecutive transitions between levels near the ground state in a double-potential well. Indeed, the observed relative peak intensities, which in the one-phonon cross-section approximation depend on the level spacing and the sample temperature, vary irregularly with increasing energy. Similar behavior has been reported for the peaks observed in the spectra of $Ca(OH)_2$ and $Mg(OH)_2$. In general, the transition between the more widely spaced lower-lying levels should be more intense in the neutron spectra than those between more closely spaced higher-lying levels because the thermal population factor favors the former. Thus, for such a potential a systematic increase in intensity with increasing peak energy would be expected. If an attempt is made to correlate the observed temperature behavior of the peaks in LiOH [assuming a potential of the type $V = V_0(1 - \cos 2\theta)$, where $V_0 = 350$ cm^{-1}, and a rotational constant $B = \hbar^2/2I = 20$ cm^{-1}] with Stern's calculations[20] for the corresponding lower-lying energy levels, the following difficulties are encountered:

1. The peak at 625 cm^{-1} would have to represent a "multilevel" or multiphonon transition from a high level to the ground state, if only to explain its energy with regard to the level separation. The intensity of the level is observed to change with temperature in accord with the differential neutron-scattering cross section for a one-phonon absorption, where $\theta \gg T$. Moreover, such an assumption does not appear in accord with the intensity of this peak at $T = 294°$K, as the rapid decrease in the neutron-scattering cross section with the increasing phonon multiplicity for $\theta \gg T$ would make such a multiphonon transition improbable compared to transitions between nearby consecutive levels,[19] which are not observed.

2. Not only is the peak at 138 cm^{-1} too broad to be associated with a transition between two consecutive levels of a highly localized potential, but it also has the shape expected for an acoustic-lattice mode. Moreover, in any attempt to account for the low-energy tail on the basis of a single-potential model it is necessary to associate this peak with unresolved transitions between closely spaced high-lying levels arising from the anharmonicity of the potential in the region of the top of the barrier. The intensity of this peak would be expected to decrease with decreasing temperature more rapidly than those of the peaks at 468 and 322 cm^{-1}. In addition, the tailing of this peak would be expected to be reduced as the temperature is lowered, and the effect of the anharmonicity is decreased. However, neither of the latter two phenomena is observed.

Based on the shape of the observed neutron spectra, it is concluded[14, 15] that the peaks result from an interaction between neutrons and lattice vibrations. The general agreement between neutron and optical data in the case of the hydroxides has led to the belief that the combination bands observed in the OH-stretching region are also due to the interaction with lattice vibrations. In the neutron and optical cases the critical points in the lattice frequency distribution appear as scattering or absorption peaks. The lowest peaks, however, are due to acoustic branches and are, therefore, not directly observable in the infrared spectra.

The neutron data[13, 15] all seem to indicate that the local uncoupled model for hydrogen motion is not suitable. It has been shown that hydrogen motion can be described in terms of waves in a layer lattice[15], and this picture is supported by neutron-inelastic-scattering data. This model can also be used to explain the optical data in a coherent way, resolving the apparent disagreement between results in the near and the far infrared. The value of inelastic neutron scattering for this problem lies once more in its sensitivity to hydrogen motion and in the fact that it is not restricted by the usual optical selection rules.

On the other hand, the good agreement between optical and neutron data indicates that investigation of lattice-vibration spectra by optical high-resolution spectroscopy of combination bands, when these exist, should also be given serious consideration.

REFERENCES

1. R. M. Hexter, *J. Chem. Phys.* **34**, 941 (1961).
2. R. A. Buchanan, E. L. Kinsey, and H. H. Caspers, *J. Chem. Phys.* **36**, 2665, (1962).
3. R. M. Hexter, *J. Chem. Phys.* **38**, 1024 (1963).
4. R. A. Buchanan and H. H. Caspers, *J. Chem. Phys.* **38**, 1025 (1963).
5. R. T. Mara and G. B. B. M. Sutherland, *J. Opt. Soc. Am.* **46**, 464 (1956).
6. R. M. Hexter and D. A. Dows, *J. Chem. Phys.* **25**, 504 (1956).
7. K. A. Wickersheim, *J. Chem. Phys.* **31**, 1 (1962).
8. S. S. Mitra, *Solid State Phys.* **13**, 1 (1962).
9. D. Krishnamurti, *Proc. Indian Acad. Sci.* **A50**, 223 (1959).
10. R. A. Buchanan, H. H. Caspers, and J. Murphy, *Appl. Opt.* **2**, 1147 (1963).
11. R. A. Buchanan, H. H. Caspers, and H. R. Marlin, *J. Chem. Phys.* **40**, 1125 (1964).
12. G. Safford, V. Brajovic, and H. Boutin, *Bull. Am. Phys. Soc.* **7**, 499 (1962).
13. G. Safford, V. Brajovic, and H. Boutin, *J. Phys. Chem. Solids* **24**, 771 (1963).
14. G. J. Safford and F. L. LoSacco, *J. Chem. Phys.* **44**, 345 (1966).
15. I. Pelah, K. Krebs, and Y. Imry, *J. Chem. Phys.* **43**, 1864 (1965).
16. D. F. Hornig, *J. Chem. Phys.* **16**, 1063 (1948).
17. H. Winston and R. S. Halford, *J. Chem. Phys.* **17**, 607 (1949).
18. S. S. Mitra, *J. Chem. Phys.* **39**, 3031 (1963).
19. L. S. Kothari and K. S. Singwi, *Solid State Phys.* **8**, 109 (1959).
20. T. E. Stern, *Proc. Roy. Soc. (London)* **A130**, 551 (1930).

8. Dynamics of Molecular Torsion and Rotation

8.1 Total-Cross-Section Measurements

Generally speaking, translational motions in molecular compounds in the same series do not differ significantly if these compounds are examined in the same condensed phase at the same temperature and pressure. On the other hand, because of differences in symmetry and electrostatic properties, rotational motions can exhibit quite different behavior. An interesting example of this situation is the motion of ammonium ions in various salts. If one assumes that translatory modes reflected by the proton motions in ammonium halides are basically the same and that rotational modes can be described by an oscillator model, then the total cross section as a function of incident neutron energy E_i can be estimated. For very long-wavelength neutrons, the dominant contribution comes from processes in which neutrons gain energy as a result of the de-excitation of the torsional mode. We shall call this the $0 \rightarrow 1$ transition, which corresponds to the one-phonon annihilation process in the lattice problem. As a first approximation, one may use a Debye model to describe the torsional mode; then it is known[1] that the resulting total cross section varies inversely as $E_i^{1/2}$:

$$\sigma_{0 \rightarrow 1}(E_i) \sim \frac{m}{M} \left(\frac{T}{\theta_D}\right)^3 \left(\frac{T}{E_i}\right)^{1/2}, \tag{8.1}$$

where M is the oscillator mass, T the temperature, and θ_D the Debye temperature. The important feature of this result is that a plot of

$\sigma(E_i)$ for very small E_i should show a straight line with slope inversely proportional to θ_D. Since θ_D is a measure of rotational barrier, the stronger the hindrance the smaller is the slope of $\sigma(E_i)$. Rush and co-workers[2-4] have performed a series of total-cross-section measurements on ammonium salts and obtained correlation of the measured slopes with barrier-to-rotation heights. This is illustrated in Figure 8.1,

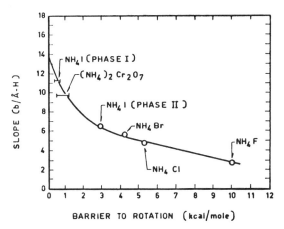

Figure 8.1. Total cross section slope versus barrier to rotation height for several ammonium compounds.

which shows that for ammonium halides there does exist a correlation; thus the slope for NH_4F is much lower than that for NH_4I, and it is well known that hindered rotation is highest for fluoride and very low for iodide.

A popular method for barrier determination in molecular solids is based on the NMR technique. For some ammonium salts, barriers are already known from NMR. Rush *et al.*[3] have shown that a calibration curve for the neutron-barrier determination can be drawn based on measured cross-section slopes and barriers derived from NMR measurements. Using this curve and the measured slope values, these authors have evaluated the unknown barriers for $(NH_4)_2CrO_4$ and NH_4CNS (4 kcal/mole) as well as that for NH_4ClO_4 (0.1–0.2 kcal/mole).

8.2   Differential-Cross-Section Measurements

Neutron scattering is sensitive to such modes as the hindered rotational modes of molecular groups in crystals, which are not readily observable by other methods. This neutron technique also offers the possibility of easily detecting the onset of free or quasi-free rotation of molecular groups. Vibrational motions (internal or external) of individual molecules or groups of atoms within the molecule can also be considered as phonons but with zero dispersion, that is, $\omega(\mathbf{q}) = $ constant. For hydrogenous groups in a molecule, the excitation or de-excitation of torsional modes provides the dominant contribution to the scattered-neutron spectrum. If the torsional modes can be described by an oscillator, the $1 \rightarrow 0$ transition between the first (thermally) excited level of the oscillator and the ground state is equivalent to the one-phonon annihilation in the case of a lattice. The methods that have been developed for evaluating the intermediate scattering function $\chi_{rot}(\kappa, t)$ for systems with only rotational degrees of freedom are limited to two extreme approximations. These correspond to describing the motions as either free rotations or small-angle torsional oscillations, the latter being obviously more relevant to the problem of inter-molecular forces. For freely rotating rigid molecules of any structure, $S(\kappa, \omega)$ and $\chi(\kappa, t)$ have been evaluated without approximation.[5] Although the exact results are available, a more widely used method has been developed by Kreiger and Nelkin.[6] The Kreiger-Nelkin approach is based on the concept of the mass tensor and, strictly speaking, is valid only when $B \ll \hbar^2 \kappa^2 / 2M \ll k_b T$, where B is a rotational constant.[7] Its popularity apparently stems from the fact that the results greatly facilitate the computation of differential and total cross sections.

From the point of view of molecular spectroscopy, the use of any free-rotation description undermines the original purpose of studying intermolecular forces. A meaningful study of rotational motions in solids and liquids should involve explicit consideration of orientation-dependent forces. As we will see in the following, only indirect or very crude analyses of the neutron data have been attempted. Obviously, much more information about molecular reorientation processes is available in the observed spectra, but it cannot be extracted without a more quantitative and fundamental treatment.

8.3   Ammonium Salts

In spite of the lack of a satisfactory theory of neutron scattering by molecular liquids and crystals, neutrons have been used by various authors as a tool for investigating the rotational dynamics of molecules.

This application may be considered complementary to the infrared or Raman techniques and also to heat-capacity measurements, from which information concerning either the rotational freedom or, in the case of hindered rotation, the torsional frequency value may be deduced. The ammonium halides have been studied by various authors over a wide temperature range covering a number of phase-transition points.[8] Because some of these points were interpreted as transitions from hindered to free rotation of the NH_4 ion, it is interesting to look at differences among the neutron spectra.[8] At higher temperatures, corresponding to phase I ($T > 185°C$ for NH_4Cl; $T > 134°C$ for NH_4BR; $T > -16°C$ for NH_4I), one obtains a broad distribution of scattered neutrons without any appearance of sharp peaks, an apparently typical situation for freely rotating NH_4 ions. At lower temperatures, on the other hand, one obtains sharp peaks, some of which may be interpreted as having been caused by torsional vibration. If the peaks are sufficiently narrow, one can calculate from their energy that of an excited level of the NH_4 ion (or the corresponding frequency in cm^{-1}). Table 8.1 presents for comparison the torsional frequencies for ammonium halides obtained by neutrons, infrared spectroscopy, and specific-heat measurements.[8]

Table 8.1. A Comparison of Ammonium Halide Torsional Frequencies Obtained by Different Methods

Substance	Torsional Frequency (cm^{-1})		
	From Neutron Measurements	From Infrared Spectra	From Specific Heat ($100°K$)
NH_4F	550	523	520
NH_4Cl	350	359	390
	355		
	388		
NH_4Br	324	311	335
	315		
	307		
NH_4I	free rotation		

It should also be mentioned that from neutron diffraction studies made for ammonium halides by Levy and Peterson,[9] it has already been concluded that the most probable motion of the NH_4 ion in the high-temperature phase (phase I) is free rotation. Thus, if one takes all these measurements (that is, neutron diffraction, neutron inelastic scattering, Raman and infrared spectroscopy, thermodynamics, and NMR) into account, one arrives at the following conclusions concerning the order and dynamics of the NH_4 ion in ammonium halides. In the low-temperature phase the NH_4 ion is ordered and performs torsional vibrations; at intermediate temperatures the NH_4 ion is disordered, but this fact does not strongly affect torsional vibrations; the high-temperature phase corresponds to a free rotation of the NH_4 ion.

The anharmonicity effect of torsional oscillations in NH_4Cl has been reported by Venkataraman et al.[10] By applying a neutron energy analyzer with much better resolution than that of an ordinary beryllium filter, the authors observed near the main torsional vibration peak at 351 cm^{-1} a second peak at 307 cm^{-1} (Figure 8.2). They interpret the main peak as being caused by the transition from the ground to the first excited torsional state, whereas the satellite is believed to be caused by a transition from the first to the second excited level. The satellite peak does not appear at temperatures as low as $135°$K, an observation that supports the interpretation given above. The potential barrier height derived from these vibrational frequencies was estimated at about 4.8 kcal/mole.

In ammonium azide (NH_4N_3), each NH_4 group is surrounded by four azide (N_3) ions at approximately tetrahedral angles. The azide (N_3) groups are linear and symmetric. The crystal structure is of the orthorhombic space group D_{2h},[7] with four molecules per unit cell. In the infrared spectrum of Dows et al.,[11] a rather intense band at 1830 cm^{-1} was assigned as a combination involving an N—H bending mode (v_4) and a torsional motion of the NH_4 ion (v_6) at 420 ± 20 cm^{-1}. The existence of this torsional frequency suggested that the rotational motion of the NH_4 group was not free but hindered by four hydrogen bonds between the nitrogen of the NH_4 ion and the closest nitrogens on the azide groups (N—N distance 2.96–2.99Å).

The NH_4 torsional frequency of 420 cm^{-1} in NH_4N_3 can be compared with corresponding frequencies of 390 cm^{-1} in NH_4Cl and 319 cm^{-1} in NH_4Br.[12] The relative magnitude of the frequencies of the three torsional modes is consistent with H-bond energies as inferred from the frequencies of the N—H stretching motions in the three ammonium salts.

The spectrum of neutrons inelastically scattered by NH_4N_3 can provide a direct observation of this torsional frequency of the NH_4

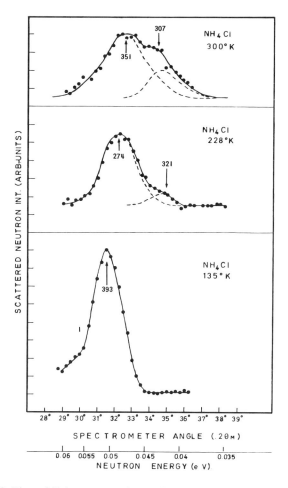

Figure 8.2. Time-of-flight spectrum of NH_4Cl in the frequency region corresponding to torsional motion of the ammonium ion as a function of temperature. A satellite peak appears as the temperature increases. A possible decomposition of the two peaks is indicated by the dotted lines. The number indicates the energy transfer in cm^{-1}.

ion.[13] It can also provide the frequencies of some lattice modes and information about the potential barrier hindering the NH_4 motions. This is indeed the case in the neutron spectrum of NH_4N_3 shown in Figure 8.3. The main peak at 400 cm^{-1} is in good agreement with the prediction from infrared data.[11] As the temperature of the sample is increased above room temperature, a shoulder at 300 cm^{-1} becomes

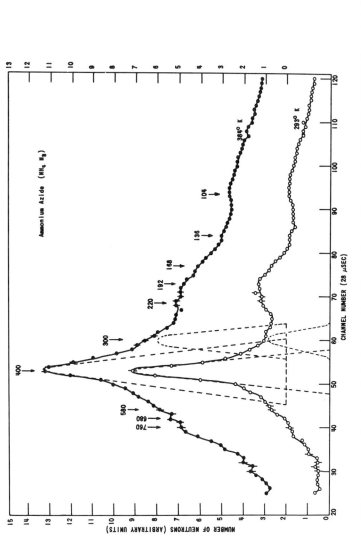

Figure 8.3. Time-of-flight spectrum of neutrons inelastically scattered by a polycrystalline sample of NH_4N_3 at 293°K (lower curve) and 384°K (upper curve). With increasing temperature a peak at 300 cm^{-1} becomes more pronounced. The energy transfers in cm^{-1} are indicated above each peak. The ordinates are proportional to the neutron scattering cross section.

more pronounced. Since no phase change or decomposition is taking place in the sample in this temperature range, the shoulder at 300 cm^{-1} is assigned to a transition between the second and the first excited states of the torsional oscillator (NH_4), while the peak at 400 cm^{-1} corresponds to a transition between the first excited state and the ground state. The variation with temperature of the ratio between the intensities of the two peaks at 300 and 400 cm^{-1} in the neutron spectrum supports this assignment. This ratio shows an increase in population of the various levels according to a Boltzmann factor. Similar transitions have been observed by neutron scattering in NH_4Cl.[10] The NH_4 ions in NH_4N_3 are therefore torsionally oscillating in a very anharmonic potential well with the first and second excited energy levels at 400 and 700 cm^{-1}, respectively, above the ground state. The symmetry of the lattice is not cubic so it is difficult to calculate exactly the height V_0 of the potential barrier, but it can be reasonably estimated at about 6 kcal/mole.

Several broad peaks at 220, 192, 168, 136, and 104 cm^{-1} are also observed in the neutron spectrum shown in Figure 8.3; they are assigned to unresolved lattice modes. The small peak at 680 cm^{-1} corresponds to the fundamental vibration v_2 of the azide ion. It is observed as a split peak at 661 and 650 cm^{-1} in the infrared spectrum. The two weak shoulders at 580 and 760 cm^{-1} may correspond to a combination band of these lattice modes with the torsional oscillation of the NH_4 group or with v_2.

Brajovic et al.[14] obtained at room temperature broad distributions of scattered neutrons for NH_4PF_6, $(NH_4)_2S_2O_8$, $NH_4K_xI_{1+x}$, and NH_4SO_3F, from which either free rotation of the NH_4 ion or only slight hindrance may be deduced. None of these compounds shows any appearance of the torsional vibration peak at room temperature (see Figure 8.4, for example). The authors attempted to check rotational freedom or hindrance in these compounds by fitting the points to a Kreiger-Nelkin-like curve with an adjusted rotational mass (taken as a parameter). A fit was interpreted to mean that rotation was free (Figure 8.4, upper curve), whereas lack of fit indicated hindrance of rotation as, for example, in the case of NH_4SO_3F or of NH_4PF_6 at low temperature (Figure 8.4, lower curve). This interpretation is not very satisfactory, however, in view of the lack of justification for applying Kreiger-Nelkin formalism to the condensed state. Rush et al.[15] have shown that in NH_4SO_3F and $(NH_4)_2S_2O_8$ the barrier to rotation of the ammonium ion is of the order of 1 kcal/mole (350 cm^{-1}).

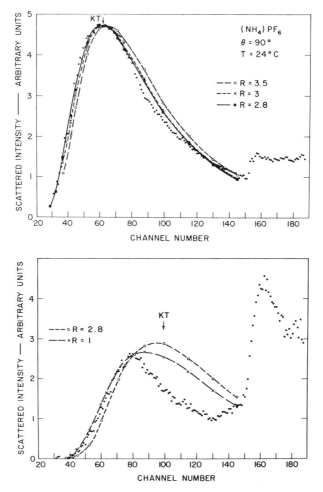

Figure 8.4. Time-of-flight spectra of neutrons scattered by a polycrystalline sample of NH_4PF_6 at (a) 297°K and (b) 93°K. The curves represent an attempt to fit the experimental data using different values of the effective rotational mass R.

8.4   Methyl Groups

Librational motions of methyl (CH_3) groups in general cannot be observed by infrared or Raman measurements in the solid state even if their observation is not forbidden by optical selection rules. This motion induces a very small change in dipole moment or polarizability, and the corresponding absorption or scattering is expected to be very small. Moreover, in most solids, it is superimposed on lattice-mode frequencies. Some of these torsional frequencies have been measured for gases by microwave spectroscopy and by far-infrared measurements. For acetaldehyde, for example, nine peaks were observed in the far-infrared spectrum of the gas, but only a weak band at 150 cm^{-1} was observed in the solid.[16] Assuming that the CH_3 group is opposed by a threefold potential barrier, the torsional energy levels become more closely spaced toward the top of the barrier (the levels of a freely rotating methyl group are such that the rotational constant is $\beta \simeq 5$ cm^{-1}). The single peak at 150 cm^{-1} in solid acetaldehyde probably corresponds to a $0 \rightarrow 1$ transition between the ground state and the first excited state of the torsional mode. A barrier height of 413 cm^{-1} or 1180 cal/mole was obtained using Herschbach's table and A levels.[17] Even in the gas phase the $0 \rightarrow 2$ transitions are already very complex and broad.

The rotational freedom of methyl groups was investigated by Rush *et al.*[4] in a way similar to that for ammonium compounds, that is, by measuring the slopes of the curves for the hydrogen scattering cross section versus neutron wavelength. The substances investigated were *o*-, *m*-, and *p*-xylene, mesitylene, and toluene; the slopes obtained were about 9.5 b/Å–H for the methyl group in *o*-xylene as compared with 11.4 b/Å for other methyl-benzenes. Using a calibration curve[4] for the CH_3 group, they estimated a barrier of about 1 kcal/mole for *o*-xylene but a much smaller one (0.1–0.4 kcal/mole) for the other methyl-benzenes. These values may be compared with those obtained by the thermodynamic method,[16] namely, 1.85 kcal/mole for *o*-xylene, 0.19 kcal/mole for mesitylene, 0.3 kcal/mole for *m*- and *p*-xylene, and 0.17 kcal/mole for toluene.

The differential neutron-scattering cross section for a number of other organic compounds was also measured by Rush and Taylor,[18] and peaks in the 100–200 cm^{-1} region were assigned to $0 \rightarrow 1$ transitions between ground states and first excited states of a Ch_3 torsional mode.

In hexamethylbenzene (HMB), this $0 \rightarrow 1$ transition of the CH_3 torsional oscillations gives rise to a distribution of frequencies in the neutron spectrum[18] peaked at about 137 cm^{-1} below 110°K. This distribution shifts to about 120 cm^{-1} above the λ point. At higher temperatures the peak disappears altogether, indicating a possible over-all rotation of the CH_3 group. The calculated barrier heights, assuming a threefold cosine potential, are, respectively, 1.07 and 1.35 kcal/mole above and below the λ point. The barrier height similarly calculated from a torsional peak at 170 cm^{-1} in the neutron spectra of durene and o-xylene[18] is larger (2 kcal/mole). It is interesting to note that in HMB the barrier heights obtained from the neutron data are much smaller than the 3- to 8-kcal/mole value obtained from thermodynamic measurements but are in agreement with those values obtained from NMR results.[18]

The neutron spectrum for acetic acid[19] (Figure 8.5a) displays a strong, broad peak at 115 cm^{-1} which has been assigned to the torsional vibration of the methyl group because of its intensity and width and also because no similar feature is observed in trifluoroacetic acid[19] (Figure 8.5b). Assuming a threefold potential barrier, the barrier height can be estimated at about 190 cm^{-1} (0.5 kcal/mole), which is only slightly higher than that in the gas phase (0.48 kcal/mole or 169 cm^{-1}).

8.5 Phenomenological Treatment of Hindered Rotations

Another approach involving a direct calculation of $\chi_{rot}(\kappa, t)$ has been used for the analysis of neutron data on water.[20, 21] A broad band of frequencies has been observed in the spectra of water and ice, and these have been interpreted as hindered rotational motions of the H_2O molecule. The first attempt to calculate χ_{rot} was made by Nelkin who used a single oscillator expression, with the mass and frequency as model parameters.[20] A more detailed study was then carried out by Yip and Osborn,[21] who postulated an effective potential of the form $\mu \cdot \epsilon$ where μ is the molecular permanent dipole moment and ϵ an effective electric field produced by neighboring dipoles. In the strong-field limit ($\mu\epsilon \gg$ rotational constant) the model gives oscillatorlike transitions similar to those in Nelkin's description. The excitation frequency turns out to be $(\mu\epsilon/I)^{1/2}$, where I is the moment of inertia. This approach is not restricted to polar systems, since by interpreting $\mu\epsilon$ as a barrier height V_0, the entire calculation is applicable to any torsional oscillation, whether external or internal in mode. The torsion frequency in this case becomes $\omega_R = (V_0/I)^{1/2}$. If the interatomic distances do not change appreciably upon deuteration and if V_0 remains roughly the same, the torsional frequency ω_R observed in a deuterated sample should be smaller. The reduction is approximately a factor of $2^{1/2}$ or less. This is a simple and effective means of identifying a torsional mode (in contrast to translatory modes which change very little) in a complex spectrum.

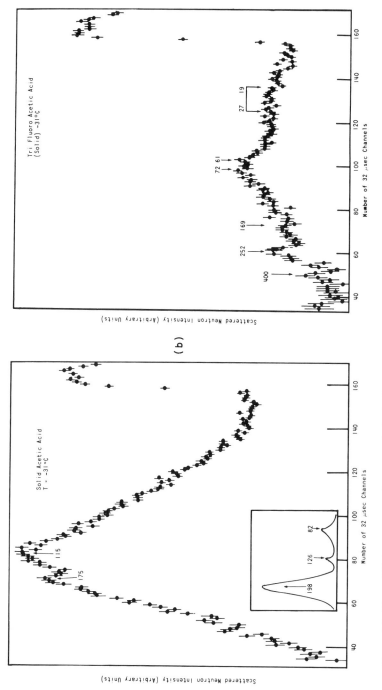

Figure 8.5. Time-of-flight spectrum of neutrons inelastically scattered by (a) solid acetic acid at 242°K and (b) solid trifluoroacetic acid at 242°K. The energy transfers are indicated in cm⁻¹ above each peak. The ordinates are proportional to the differential scattering cross section. In the insert the far infrared spectrum of acetic acid is shown for comparison.

REFERENCES

1. L. S. Kothari and K. S. Singwi, *Solid State Phys.* **8**, 110 (1959).
2. J. J. Rush, T. I. Taylor, and W. W. Havens, Jr., *Phys. Rev. Letters* **5**, 507 (1960).
3. J. J. Rush, T. I. Taylor, and W. W. Havens, *J. Chem. Phys.* **35**, 2265 (1961).
4. J. J. Rush, T. I. Taylor, and W. W. Havens, *J. Chem. Phys.* **37**, 234 (1962).
5. A. Rahman, *J. Nucl. Energy* **A13**, 128 (1961).
6. T. J. Krieger and M. S. Nelkin, *Phys. Rev.* **106**, 290 (1957).
7. J. A. Janik and A. Kowalska, in *Thermal Neutron Scattering,* edited by P. A. Egelstaff (Academic Press, Inc., New York, 1965), Chaps. 9 and 10; see also Janik in Ref. 8.
8. J. A. Janik, in *Inelastic Scattering of Neutrons in Solids and Liquids* (International Atomic Energy Agency, Vienna, 1965), Vol. 2, p. 243.
9. H. A. Levy and S. W. Peterson, *Phys. Rev.* **86**, 766 (1952).
10. G. Venkataraman, K. U. Deniz, P. K. Iyengar, P. R. Vijayaraghavan, and A. P. Roy, *Solid State Commun.* **2**, 17 (1964).
11. D. A. Dows, E. Whittle, and G. C. Pimentel, *J. Chem. Phys.* **23**, 1475 (1955).
12. E. D. Wagner and D. F. Hornig, *J. Chem. Phys.* **18**, 296 (1950).
13. H. Boutin, S. Trevino, and H. J. Prask, *J. Chem. Phys.* **45**, 401 (1966).
14. V. Brajovic, H. Boutin, G. J. Safford, and H. Palevsky, *J. Phys. Chem. Solids,* **24**, 617 (1963).
15. J. J. Rush, T. I. Taylor, and W. W. Havens, *Nucl. Sc. Eng.* **14**, 339 (1962).
16. C. A. Wulf, *J. Chem. Phys.* **39**, 1227 (1963).
17. W. G. Fateley and F. A. Miller, *Spectrochim. Acta* **17**, 857 (1961).
18. J. J. Rush and T. I. Taylor, *J. Chem. Phys.* **44**, 2746 (1966).
19. H. Boutin and C. V. Berney, unpublished results.
20. M. S. Nelkin, *Phys. Rev.* **119**, 741 (1960).
21. S. Yip and R. K. Osborn, *Phys. Rev.* **130**, 1860 (1963).

9. Ice, Water, Aqueous Solutions, and Salt Hydrates

9.1 Ice

It was shown in Chapter 3 that in one-phonon incoherent scattering by monatomic cubic crystals the energy spectrum of scattered neutrons is directly proportional to the density of phonon states. This relation makes it possible to measure the phonon frequency distribution function $g(\omega)$ in certain crystals (vanadium, for example). For crystals with more than one atom per unit cell, the neutron spectrum is determined by a frequency function $G(\omega)$ which differs fundamentally from $g(\omega)$. Whereas $g(\omega)$ can be defined in terms of the dispersion relation, $G(\omega)$ involves both the dispersion relation and the polarization vectors of normal modes.

A method for determining the low-frequency portion of $g(\omega)$ for molecular solids has been outlined in Chapter 3. It was assumed that in this region $g(\omega)$ has contributions from motions associated with center-of-mass translations and molecular rotations (or librations). It was also shown that the neutron-scattering data depend on an effective frequency distribution $G(\omega)$ which is quite simply related to an approximate $g(\omega)$. Under favorable conditions this approach results in an essentially direct experimental determination of $g(\omega)$.

In molecular systems the intermolecular interactions are usually much weaker than the intramolecular forces, and it is therefore reasonable to separate the external modes (translations and rotations) from the internal modes (intramolecular vibrations). The external modes

143

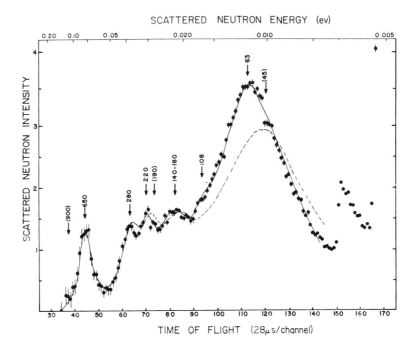

Figure 9.1. Time-of-flight spectrum of neutrons inelastically scattered through 65° from hexagonal ice at 150°K. The indicated vibrational energies are in cm^{-1}. The curves are discussed in the text.

are of interest in low-energy neutron experiments because excitations of internal states are quite unlikely for most systems. To make the calculation tractable, it is further assumed that translation-rotation couplings can be neglected. This is admittedly a rather drastic approximation, and it constitutes the most serious source of error in the determination of $g(\omega)$.

Using the same assumptions Hahn[1] has also derived an expression for the incoherent cross section for molecular crystals and has introduced different frequency functions for translations and librations. His analysis is quite formal and bears close resemblance to lattice-dynamics calculations. In both cases an enhancement factor for the librational effects in the neutron spectrum was found. It appears, therefore, that neutron data alone are not sufficient to derive $g(\omega)$. However, this difficulty can be partly avoided if either the translation or the libration component is known *a priori* or if they can be assumed to occur in distinct frequency regions. This method has been used by Prask *et al.*[2] to obtain a spectrum of molecular frequencies in ice; they found that the separation of translation and libration frequencies was in this case

Table 9.1. External Modes of Vibration in Hexagonal H_2O Ice (Frequencies, cm^{-1})

Raman (R) and Infrared (IR) Data Compiled by Ockman[a]	Recent Far-Infrared Data[b]			Neutron Data	
	100°K	168°K	193°K	Previous[c]	Prask *et al.*[d]
53 (R)	(67)		65	53	~45
					63
97 (R)	103	~100	~100		108
122 (R)			~140		140
160 (IR)	164	160	158		160
177 (R)					
193 (R)	190	184	185	~200	~190
212 (R)					
232 (R)	229	222	222		220
252 (R)		~260			
272 (R)	~275				275
294 (R)	~305	~300			
457 (R)					
516 (R)					
540 (IR)	555				
600 (R)					
				610	
660 (IR)	660			630	650
				640	
800 (IR)	770				
	840			850	~900
	900				

[a] Review by N. Ockman, *Advan. Phys.* **7**, 199 (1958).

[b] W. Bagdade, *Proceedings of the Mid-American Symposium on Spectroscopy*, Chicago, May, 1967; J. E. Bertie and E. Whalley, *J. Chem. Phys.* **46**, 1271 (1967).

[c] D. J. Hughes, H. Palevsky, W. Kley, and E. Tunkelo, *Phys. Rev.* **119**, 872 (1960); R. L. Stearns, H. R. Meuther, and H. Palevsky, *Bull. Am. Phys. Soc.* **6**, 71 (1961). K. E. Larsson and U. Dahlberg, in *Inelastic Scattering of Neutrons in Solids and Liquids* (International Atomic Energy Agency, Vienna, 1963), Vol. 1, p. 317; *Reactor Sci. Tech.* **16**, 81 (1962).

[d] H. Prask, H. Boutin, and S. Yip, *Proceedings of the Mid-American Symposium on Spectroscopy*, Chicago, May, 1967 (to be published); *J. Chem. Phys.* **48**, 3367 (1968).

justified, and their results generally agreed with some simplified calculations of the dispersion curves of ice by Forslind.[3]

The neutron time-of-flight spectrum of hexagonal ice obtained by these authors at 150°K and at a scattering angle of 65° is shown by the points in Figure 9.1. The indicated vibration frequencies were assigned after instrumental resolution, thermal population, and incident beam width were taken into account. The dotted lines correspond to calculated time-of-flight spectra and will be discussed in detail later in this section. The results are in good agreement with the neutron measurements of several other investigators.[4] A comparison of intermolecular vibration frequencies as observed in Raman, infrared, and neutron-inelastic-scattering measurements is shown in Table 9.1. The optical frequencies compiled by Ockman[5] are presented in this table along with more recent data.[6]

In general, good agreement between vibration frequencies obtained by optical techniques and those obtained by neutron inelastic scattering is not expected since optical measurements give $k = 0$ values and are governed by symmetry-dependent selection rules, whereas neutron measurements show maxima and minima in the density of states. Ice (Ih), however, is a disordered crystal so that, according to the theory of Whalley and Bertie,[7] both Raman and infrared spectra should also show features of the density of states. The agreement between the optical and neutron measurements in the translational region ($\hbar\omega \leq 350 \text{ cm}^{-1}$) is indeed quite good.

The use of Equation 3.57 to obtain an approximate $G(\omega)$ directly from the data is, in principle, straightforward. However, energy-gain neutron-scattering spectra obtained with the entire beryllium-filtered spectrum as the incident beam are appreciably distorted, particularly for low-energy transfers ($E_{\text{final}} \leq 300 \text{ cm}^{-1}$). An attempt has been made to correct for this effect and also for that of broadening due to instrumental resolution. This was done by calculating the time-of-flight spectrum from $G^0(\omega)$ [setting $\exp(-2W) = 1$] by means of Equation 3.57 and then folding in the incident neutron distribution and the calculated instrumental resolution. The incident spectrum was obtained by measuring the elastically scattered neutron distribution from vanadium which has a high Debye temperature and therefore gives an undistorted incident distribution.

The initial $G^0(\omega)$ chosen was that obtained by directly inverting the data of Figure 9.1 by means of Equation 3.57. When this initial distribution is used to recalculate the time-of-flight spectrum with instrumental effects included, the dashed line shown in Figure 9.1 is obtained. The final calculated time-of-flight spectrum, shown as a solid line in Figure 9.1, was obtained by adjusting the initial distribution until reasonably good agreement with the measured data was obtained. The corresponding $G^0(\omega)$ is shown in Figure 9.2. It should be mentioned

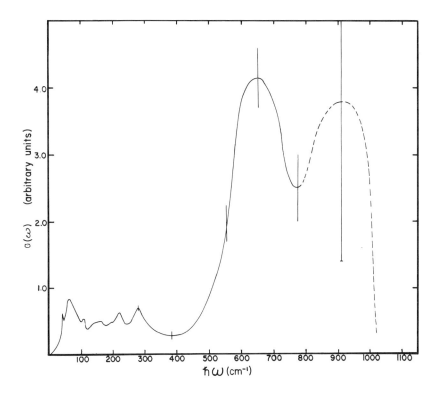

Figure 9.2. The effective frequency distribution $G(\omega)$ of hexagonal ice $[\exp(-2W)=1]$.

that although the uncertainties in the librational portion of the spectrum ($E \gtrsim 400$ cm^{-1}) are quite large, transitions at 630 and 850 cm^{-1} have been observed in ice by Woods *et al.*[4] in an energy-loss neutron-scattering experiment.

An effect that should be taken into account in analyzing neutron-scattering data is that of multiple scattering in the sample. Slaggie[8] has shown that multiple scattering is appreciable in neutron-scattering measurements on room-temperature water. The calculation has not yet been performed for ice and no correction for multiple scattering was made by Prask *et al.*[2] In view of the uncertainties mentioned above, they felt that too detailed an analysis of the data was unwarranted, so the following procedure was used. The Debye-Waller factor was calculated, where it was assumed that $M_R = 2$.[9, 10] A Gaussian curve was fit to the librational peak at 660 cm^{-1} (half-width at half-maximum = 200 cm^{-1}) and extended to lower energies, thus defining a separation into translational and librational modes. An iterative procedure was then followed and the Debye-Waller factor calculated with the final requirement being $G(\omega) \exp(-2W) = G^0(\omega)$ [$G^0(\omega)$ is shown in Figure 9.2], the initial $\exp(-2W)$ having been calculated from $G^0(\omega)$ itself. The above relation assures that $G(\omega)$, when inserted into Equation 3.57, gives the measured time-of-flight spectrum when the energy-transfer-dependent Debye-Waller factor is used. The effective frequency distribution $G(\omega)$ is shown in Figure 9.3. The Debye-Waller factor is determined by the amplitude of molecular vibration, that is, $2W = \kappa^2 \overline{u^2}$; $(\overline{u^2})^{1/2}$ is found to be 0.188Å ($T = 150°$K) from the above analysis. Peterson and Levy determined by means of neutron diffraction

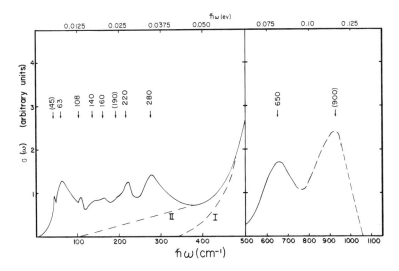

Figure 9.3. The effective frequency distribution [exp $(-2W)$ included] of hexagonal ice. Curve I corresponds to a separation wherein a Gaussian is fitted to the librational peak at 650 cm^{-1} (full width at half-maximum = 200 cm^{-1}); Curve II corresponds to a linear extrapolation of the librational distribution.

the average rotational mean-square values for oxygen and deuterium in D_2O ice to be 0.138 and 0.167Å at 123°K and 0.173 and 0.201Å at 223°K, respectively.[11] The rotational mean-square amplitude of hydrogen in H_2O ice has been measured to be 0.18 ± 0.02Å at 4°K.[11] Leadbetter,[12] by analyzing specific-heat data, has inferred $(\overline{u^2})^{\frac{1}{2}}$ values for oxygen of 0.160 and 0.195Å, respectively.

Since there is no criterion for making the separation into translational and librational modes, even in this "favorable" case, the specific heat was calculated for the Gaussian librational distribution previously described and for the somewhat arbitrary linear extrapolation of the librational modes also indicated in Figure 9.3. These two separations are taken to represent the limits of reasonable choice consistent with both neutron and optical isotope shift measurements on D_2O and H_2O ice. The translational and librational distributions for the calculation of $C_v(T)$ should arise from equal total degrees of freedom. Therefore, the final normalization of the effective distribution was performed by requiring that $\int g_T(\omega)\,d\omega = \int g_L(\omega)\,d\omega$, which implicitly assumes that all external degrees of freedom contribute to the observed spectrum.

It should be pointed out that within the limitations of the simple theoretical approach presented here, if the effective rotational mass is known, an iterative procedure can be used to determine the Debye-Waller factor and the multiphonon contributions to the measured spectrum. The latter is possible because the intermediate scattering function given in Equation 3.56 can be expanded exactly as in the monatomic case[13] and expressions for the various multiphonon terms obtained.

Figure 9.4 shows a comparison of the measured values of $C_v(T)$ for 3–100°K with values calculated from the two separations just described. The measured $C_v(T)$ values were obtained from actual $C_p(T)$ measurements through the use of coefficient of cubic expansion and adiabatic compressibility values compiled by Leadbetter.[12] The contribution to $C_v(T)$ of the intramolecular modes is negligible even at 273°K; the anharmonic contribution, $C_v^{\mathrm{anh}}(T)$, has been calculated by Leadbetter and is negligible for 0–100°K. Implicit in this comparison is the assumption that the frequency distribution obtained at 150°K is essentially the same as that at 0°K. The results of Leadbetter indicate that the average translation and libration frequency shift should be less than 2 percent in going from 0°K to 150°K, a deviation which for the purpose of the present analysis is quite small.

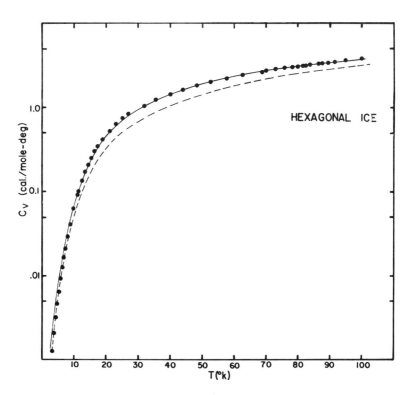

Figure 9.4. The specific heat of ice, 0–100°K. The solid points are the measured values. The dashed line is $C_v(T)$ calculated using separation I of Figure 9.3; the solid line the calculated $C_v(T)$ using separation II of Figure 9.3.

It is clear from Figure 9.4 that for both separations the shape of the calculated $C_v(T)$ agrees well with the measured values, and the linear extrapolation shows excellent absolute agreement. This suggests that the low-frequency part of the neutron frequency distribution is quite close to the true thermodynamic distribution.

Leadbetter has analyzed the low-temperature portion of the specific-heat curve as a power expansion in temperature. The coefficients obtained are in turn simply related to the coefficients of a power expansion in frequency of the density of states. A comparison of his results with the neutron results along with a Debye distribution ($\theta_D = 220°K$) are shown in Figure 9.5. The agreement is indeed quite good. Leadbetter's analysis suggests an Einstein mode at 48 cm^{-1} containing 7 percent of the translational degrees of freedom. Both the

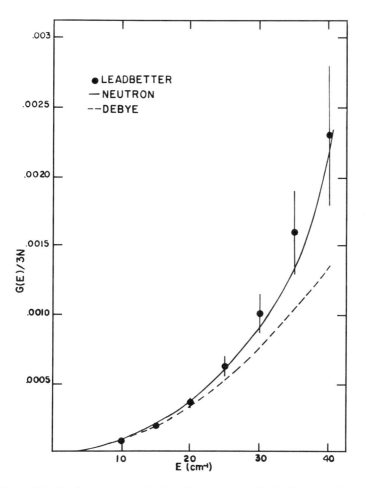

Figure 9.5. The frequency distribution of hexagonal ice for 0–40 cm^{-1}. The points and associated uncertainties are from Ref. 6. The dashed line is a Debye frequency distribution ($\theta_D = 220°$K). The solid line is taken from the work of Prask *et al.*[2]

neutron and the far-infrared[13] measurements show that the main low-energy peak occurs at about 63 cm^{-1}.

The specific heat over the entire temperature range 0–273°K is shown in Figure 9.6. The measured C_p are from the data of Flubacher *et al.* (2–20°K)[14] and Giauque and Stout (20–267°K).[15] The $C_v(T)$ shown are obtained as described earlier and the $C_v(T)^{\text{anh}}$ are from the analysis of Leadbetter. The absolute $C_v(T)$ calculated from both the

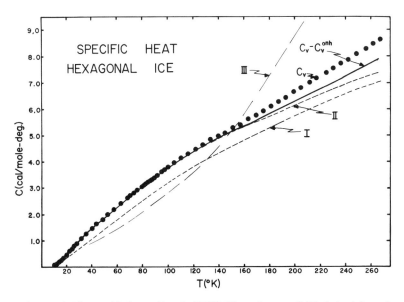

Figure 9.6. The specific heat of ice, 0–273°K. The points are $C_v(T)$ derived from the measured $C_p(T)$. The solid line corresponds to $C_v^v(T) - C_v^{anh}(T)$. Curves I and II are, respectively, $C_v(T)$ calculated from separations I and II of Figure 9.3 (see text). Curve III is $C_v(T)$ calculated directly from $G(\omega)$ without taking into account the effective rotational mass and is normalized to the measured $C_v(T)$ value at 150°K.

Gaussian and the linear extrapolation separations of $G(\omega)$ are shown. Also shown are $C_v(T)$ calculated from the "monatomic" neutron frequency distribution $[G(\omega)]$ normalized at 150°K to the measured C_v (150°K). In view of the uncertainties in this $G(\omega)$ value, the agreement is reasonably good.

Finally, Leadbetter has determined the moments of the frequency distribution of ice from his detailed analysis of the specific heat.[12] Table 9.2 shows a comparison of Leadbetter's values with the reduced moments of the translational distributions of the two separations of $G(\omega)$ considered here. The agreement between the linear extrapolation separation and Leadbetter's values is quite good. The large uncertainties in the librational distribution and the effects not yet corrected for do not allow a meaningful comparison of libration moments.

The distributions of translation and libration modes appear in G with different intensity factors. The presence of an enhancement factor for librations leads to the general conclusion that the inelastic portion of the observed incoherent neutron spectrum is, in general, dominated

Table 9.2. Reduced Moments of $g_T(\omega)$

| Moment | Neutron Data[a] | | Leadbetter Data[b] |
	I^c	II^d	C_v
$(\mu^2)_T^{1/2}$	226	190	190 ± 3
$(\mu^4)_T^{1/4}$	256	218	212 ± 5
$(\mu^6)_T^{1/6}$	275	236	225 ± 7

[a] H. Prask, H. Boutin, and S. Yip, *Proceedings of the Mid-American Symposium on Spectroscopy,* Chicago, May, 1967 (to be published).
[b] A. J. Leadbetter, *Proc. Roy. Soc. (London)* **287A**, 403 (1965).
[c] Gaussian libration separation.
[d] Linear extrapolation separation.

by rotational dynamics. In the approach here described, the effective rotation mass M_R is, strictly speaking, a κ-dependent quantity; this implies that the intermediate scattering function for a purely rotational system is not identical to the oscillator result even in the torsional limit. Approximate expressions for estimating M_R have been given by Hahn[1] and by Prask *et al.*,[2] but in practice these may require further adjustment to give optimum results.

The relation between neutron cross section and G does not permit an experimental determination of the actual frequency spectrum $g(\omega)$, even in the one-phonon approximation. To remove the enhancement factor (M/M_R), a knowledge of individual distributions g_T and g_R is needed. Since for most systems these are not accurately known, if they are known at all, a more practical procedure is to assume that translation and libration frequencies occur in separate regions. This is clearly a more restricted method, but it is probably reasonable so long as the atomic masses involved in rotation are small compared to the entire molecular mass. Because of the relatively strong intermolecular forces due to hydrogen bonds and because of the low moments of inertia of the molecule, the derived $G(\omega)$ is at least sufficiently reliable to yield thermodynamic calculations in good agreement with measurements.

9.2 Liquid Water

In the case of liquid water, one can extract a frequency distribution from neutron data, using the same approximations as those made for ice.[16] This is shown in Figure 9.7. In the case of H_2O vapor, the bondings become very weak or nonexistent, and the molecules undergo free

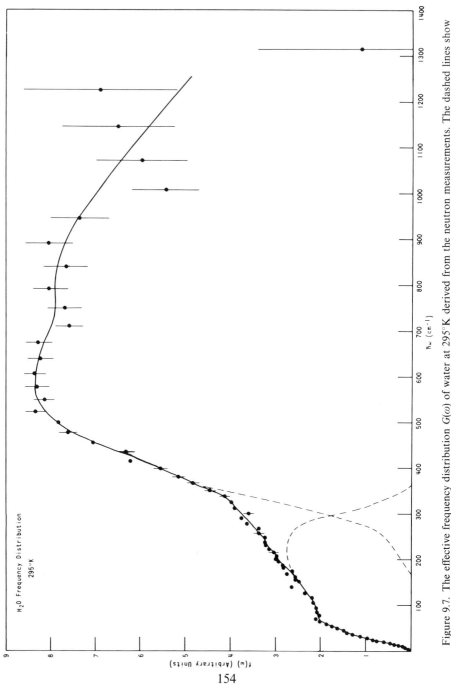

Figure 9.7. The effective frequency distribution $G(\omega)$ of water at 295°K derived from the neutron measurements. The dashed lines show a possible division into translational and rotational parts.

rotation. For the reason presented in the previous chapter, the neutron spectrum under these circumstances displays a very broad distribution of frequencies. The data of Hughes *et al.*[4] are shown in Figure 9.8. The

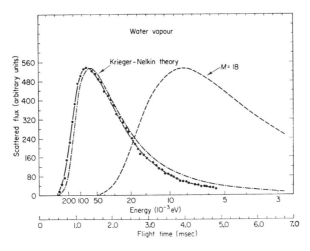

Figure 9.8. The differential neutron scattering cross section of water vapor obtained experimentally (black dots). The dotted lines correspond to theoretical curves obtained by the Krieger-Nelkin theory using different values for the effective mass M of the water molecule.

possibility of determining the structural parameters of liquid water from cold-neutron-scattering data has been discussed recently by A. Szkatula and A. Fuliuski.[17] They found that the dynamical and associational parts of the thermodynamic functions—heat capacity C_v, for example—can be measured from neutron data. This permits the calculation of the fraction of hydrogen bonds still existing in liquid water, along with their energy. On the basis of currently available neutron data,[4] the dynamical and associational parts of C_v are available for a temperature range between $0°C$ and $17°C$. The values of both parts of the heat capacity of water and of the heat of melting of ice at $0°C$ are estimated to be

$$C_v^{dyn} \approx 10.5 \text{ cal/mole/degree}, \quad L^{dyn} \approx 210 \text{ cal/mole},$$
$$C_v^{assoc} \approx 7.5 \text{ cal/mole/degree}, \quad L^{assoc} \approx 1210 \text{ cal/mole}.$$

The above data lead to the following approximate values for the energy of H-bond formation (Q) and for the number of moles of H bonds existing in water (n_0) at $T = 0°C$:

$$Q \simeq 2 \text{ kcal/mole}, n_0 = 1.4 \text{ mole}.$$

The latter value implies that during the melting of ice about 30 percent of the hydrogen bonds are broken.

9.3   Molecular Motion in Aqueous Solutions

Neutron-scattering studies of pure water have been extensive; however, only a few studies of aqueous solutions have been reported. Stiller[18] has examined the scattered spectrum at very low-energy transfers ($h\omega \leq 10 \text{ cm}^{-1}$) in H_2O, in liquid and solid aqueous HF solutions, and in solid solutions of NH_4F. His results indicate the presence of discrete low-energy transitions in each case, which are, at least for the HF solutions, consistent with a proposed phenomeno-logical description of tunnel transitions between hydrogen-bonding sites.

G. J. Safford and A. W. Naumann[19] extensively investigated the inelastic portions of the neutron spectra of salt solutions. They found that at concentrations of 4.6 M and below many of the frequencies observed in the spectra of pure water also appear in those of salt solutions. The sharpness and intensities of these lines are not the same for all solutions, but the differences observed between solutions or between solutions and pure water are not large. This indicates that even though H_2O-ion interaction occurs, the long-range disorder of the associated units is not strongly altered by the presence of ions at these concentrations. At the same concentrations a spectral intensifica-tion is observed in the region below the torsional maximum (400 to 200 cm^{-1}). This maximum is broad and shows little if any resolution of new frequencies. It occurs in the region where cation-H_2O stretching frequencies are expected. In view of the structural disorder of the associated units discussed above, the broadness of this additional component suggests that there exists a variety of coordination between the cations and the water molecules. At higher concentrations the spectra of a number of salts (KF, $AlCl_3$, $LiNO_3$, LiCl, and $MgCl_2$) approach those of the corresponding solid salt hydrates. Thus, at high concentrations the coordination of water molecules to ions becomes more distinctive and approaches that found in the solid salt hydrates.

With the introduction of electrolytes the quasi-elastic region of the water spectrum in the vicinity of the elastic peak shows more pro-nounced changes than are seen in the inelastic region. It has been shown,[19] for example, that for cations of essentially the same radius but different ionic charge the area under the quasi-elastic peak increases regularly with increasing charge; that is, $NaCl < CaCl_2 < LaCl_3$. It has also been shown that for cations of a given charge the size of the quasi-elastic peak increases regularly with decreasing ionic radius; that is, $CsCl < KCl < NaCl < LiCl, SrCl_2 < CaCl_2 < MgCl_2, LaCl_3 < AlCl_3$. For salts with a common cation but differing anions the elastic peak increases regularly with decreasing ionic radius; that is, $KCNS < KI < KBr < KCl < KF$. Moreover, the variations observed for these monovalent anions are considerably larger than those observed for the monovalent cations.

Safford and Naumann also found that there is a regular increase in the size of the quasi-elastic peak with salt concentration. These changes in the quasi-elastic region are due to an increase in the average Debye temperature or, equivalently, to a decrease in the average vibrational amplitude of water molecules because of H_2O-ion interactions that are stronger than the H_2O—H_2O hydrogen bonding of pure water. However, as shown in Table 9.3, for certain salts the quasi-elastic peak becomes sharper than for water, while for others it appears the same or

Table 9.3. Effects of Salts on the Quasi-Elastic Peak of Water[a]

Large increase in area, large decrease in width	Moderate increase in area, slight decrease in width	Small increase in area, slight increase in width	Approximately the same as pure water
$AlCl_3$(3.5 M) $LaCl_3$(3.5 M) $MgCl_2$(3.5 M) $CaCl_2$(3.5 M) $SrCl_2$(3.5 M)	$MgCl_2$(1 M)		
KF(17 M) $LiCl$(18.5 M)	$KCNS$(18.5 M) KF(4.6 M) $LiCl$(4.6 M) $NaCl$(4.6 M)		$KCNS$(4.6 M)
			$NaCl$(3.5 M)
		KCl(4.6 M) $CsCl$(4.6 M)	
	$NaClO_4$(4.6 M) $Mg(ClO_4)_2$(4.6 M)		
		KBr(4.6 M)	KI(4.6 M)

[a] Peak descriptions are based on a comparison of difference curves obtained by the subtraction of the spectrum of pure water from those of the salt solutions.

broader. The former condition indicates a decrease in diffusive motion or an increase in the residence time of water molecules strongly co-ordinated with ions. The latter is very probably caused by a further reduction in the already poor ordering of the associated units. A quantitative study of the changes in the diffusive broadening of the quasi-elastic regions would provide more exact information on ion-induced changes in the diffusive kinetics of water molecules.

Inelastic-neutron-scattering studies have also been performed by Prask and Boutin[16] on aqueous solutions of electrolytes and organic molecules. Figure 9.9 shows neutron spectra obtained for 3.4 M solutions of halides of potassium and magnesium at room temperature. For comparison, dashed curves corresponding to the spectrum of pure water are shown, the indicated transitions being those in water. Unlike the results of Raman measurements,[20] spectral changes appear to be primarily due to cation rather than anion differences. This is illustrated more clearly in Figure 9.10, which presents the ratios of intensities per hydrogen of pure H_2O in solution as a function of time-of-flight or neutron energy. It is apparent that the ion effects manifest themselves almost entirely in the lower-frequency portion ($h\omega \leq 100$ cm^{-1}) of the spectra. If water is considered quasi-crystalline, this region corresponds to intermolecular translation vibrations of the "lattice" by analogy with the ice spectrum, where such vibrations have been shown to occur from 0 to 300 cm^{-1} (Figure 9.1). Small, highly charged cations cause damping of the longest-wavelength "phonons" of water, whereas large, monovalent anions have little effect. Harris and O'Konski[21] have proposed a model for ion-H_2O molecule interaction to explain results of dielectric-constant measurements in aqueous electrolyte solutions as shown in Figure 9.11, where because of cation interaction the H_2O is not free to rotate. This model has also been used to interpret NMR results for several solutions.[22]

Such an approach appears consistent with the neutron results of Prask and Boutin.[16] Because of the rotational freedom of the co-ordinated water, large, monovalent anions may replace a water molecule in the lattice without seriously disrupting the hydrogen-bond network. The smaller, divalent cations, on the other hand, bind co-ordinated water molecules more strongly and hinder their rotational motion; noncoordinated water molecules therefore cannot form hydrogen bonds to these without disrupting the "lattice" in the neighborhood of the ions. The addition of the cations should have the greater effect on long-range order and consequently on the long-

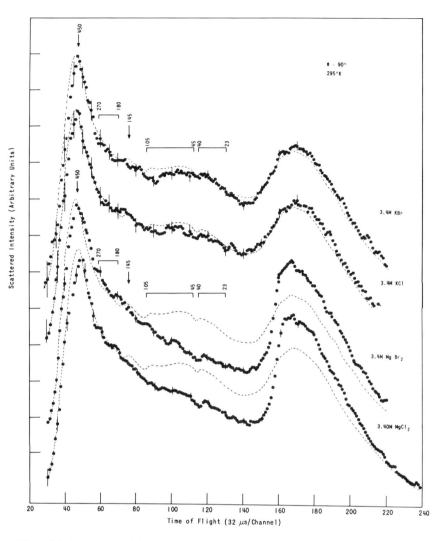

Figure 9.9. Neutron partial differential cross sections versus time-of-flight for 3.4M aqueous solutions of KBr, KCl, MgBr$_2$, and MgCl$_2$ at 295°K. The dashed curves correspond to the spectrum of pure water and the indicated energies (in cm^{-1}) correspond to transitions in water.

wavelength "phonon" modes of the quasi-lattice, with charge density being the determining factor (Mg^{++} > Li$^+$ ⩾ Ca^{++} > K$^+$). The order Mg^{++} > Ca^{++} = Li$^+$ > K$^+$ was experimentally observed. This cation-effect ordering is also evidenced by the sharpening of the

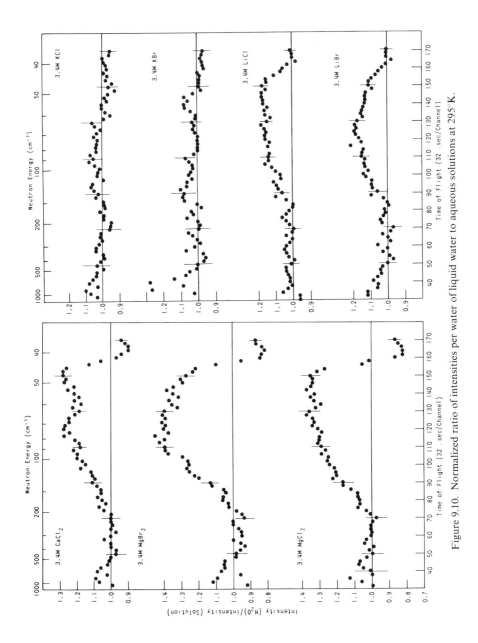

Figure 9.10. Normalized ratio of intensities per water of liquid water to aqueous solutions at 295°K.

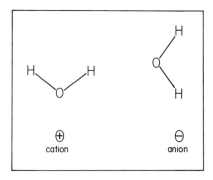

Fig. 9.11. Possible orientation of a water molecule with respect to neighboring anion or cation.

elastic peaks in the neutron spectra. Table 9.4 shows the broadening for the various electrolytes, where a Lorentzian energy distribution corrected for instrumental resolution has been assumed. The full-width at half-maximum, ΔE, also decreases in the order K^+, Li^+, Ca^{++}, Mg^{++}. If water molecules form several hydration shells around each ion in solution, as postulated by Frank and Wen,[23] the neutron spectra might be expected to consist of a superposition of vibrational motions corresponding to the various species of H_2O molecules. The changes observed for the relatively high concentrations ($\sim$ 15 H_2O/cation) in the measurements of Prask and Boutin,[16] however, show no evidence to favor this description in either the elastic- or inelastic-scattering region of the spectrum. The neutron studies of electrolyte solutions also indicate that the structural changes produced by adding ions to water are different in nature from the bonding changes produced by varying the temperature of pure water; thus the idea of a "structural tempera-

Table 9.4. Broadening of the Quasi-Elastic Peaks for Various Electrolytes

Solution	$\Delta E(\text{cm}^{-1})$
H_2O	5.0 ± 1.5
3.4-M KCl	5.0 ± 1.5
3.4-M KBr	5.0 ± 1.5
3.4-M LiCl	3.5 ± 1.0
3.4-M LiBr	3.5 ± 1.0
3.4-M $CaCl_2$	3.0 ± 1.0
3.4-M $MgBr_2$	2.5 ± 1.0
1.7-M $MgCl_2$	3.5 ± 1.0
3.4-M $MgCl_2$	2.0 ± 1.0

ture" seems unfounded. This conclusion is in agreement with the earlier results of Kujumzelis[24] and Berqvist and Forslind.[25]

Neutron-scattering measurements have also been made on aqueous solutions of some deuterated organic solutes, among them 3.7 M deuterated acetone and 3.9 M deuterated dimethylsulfoxide.[16] In all cases spectra are almost identical to that of pure water except for a marked sharpening of the quasi-elastic peak. This result indicates a decreased mobility of H_2O molecules in solution and is not inconsistent with the suggestion of Frank and Wen[23] that nonpolar solvents are "structure"-making in their effect.

9.4   Molecular Motions of H_2O Molecules in Salt Hydrates

The strength of the bonds formed with neighboring atoms or molecules, the proximity of cations, and the environment all influence the librational and translational motion of H_2O molecules. Preliminary information obtained for well-characterized systems is essential to an interpretation of neutron spectra for molecules adsorbed on surfaces where little structural data is available. It then becomes possible to characterize an adsorbed molecule by its solid-, liquid-, or gas-like properties, depending, for example, on the relative amount of time it may spend vibrating around an equilibrium position and diffusing to another site. Vibrational motions can be obtained for each site if it is assumed, of course, that they have been identified by means of other techniques.

The vibration frequencies (librational and translational) of H_2O molecules have been studied in a large variety of known configurations and environments. The hope has been that these frequencies would characterize some distortion of H_2O molecules so that analogies could be drawn with H_2O molecules adsorbed on surfaces and interpretations made on the basis of their neutron spectra.[26]

Molecular motions have been studied in crystal hydrates and frequencies obtained for cases where neither of the lone electron pairs on the oxygen of the H_2O molecule is specifically directed ($K_2C_2O_4$ H_2O), where the lone pairs are directed toward monovalent metal ions ($KF \cdot 2H_2O$), and where the H_2O molecule has a single orientation $[(BaClO_3)_2 \cdot H_2O]$ or two distinct and equally probable orientations ($Li_2SO_4 \cdot H_2O$).[26] The effects of various cations and anions have also been investigated for a series of hydrates in which the H_2O is H-bonded to the same anion Cl^- but on the sphere of coordination of different cations (Al^{+++}, Cr^{+++}, Sr^{++}, Fe^{++}, Mg^{++}, Co^{++}, Ni^{++}) and for Al salts with different anions (NO_3^-, SO_4^{--}, Cl^-). The neutron spectrum is in every case characteristic of the H_2O molecule and consists essentially of a torsional band in the 400–700 cm^{-1} region (usually a composite of several frequencies) and some translational bands in the 100–300 cm^{-1} region which are assigned to metal-OH_2 stretching or to stretching of the hydrogen bond between two H_2O molecules or between an H_2O molecule and a neighboring anion.

As an example, the neutron spectrum of $K_2C_2O_4 \cdot H_2O$ is shown in Figure 9.12.[26] The torsional motions of the H_2O molecule around its

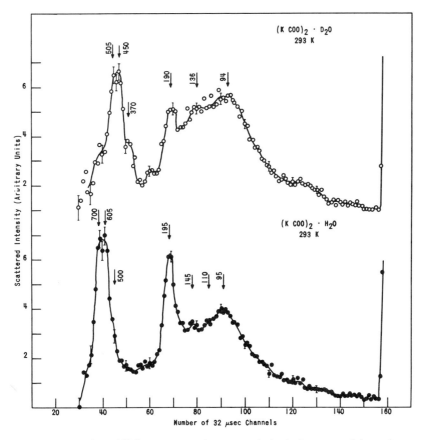

Figure 9.12. Time-of-flight spectrum of neutrons inelastically scattered by polycrystalline samples of K(COO)₂·H₂O and K(COO)₂·D₂O. The energy transfers in cm^{-1} are indicated above each peak.

axis of symmetry can be identified by their shift to lower frequencies upon deuteration; the translational bands are not affected by deuteration. Three such torsional frequencies are expected for each species of H_2O molecules. Only in a few cases can all the optically active frequencies be observed in the infrared spectrum. In $BaCl_2 \cdot 2H_2O$, H_2O molecules can be hydrogen-bonded either to two chlorine atoms (type I) or to a chlorine atom and an oxygen from a neighboring H_2O

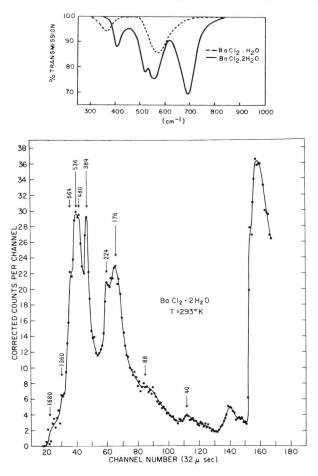

Figure 9.13. Time-of-flight spectrum of neutrons inelastically scattered by polycrystalline samples of $BaCl_2 \cdot 2H_2O$. The energy transfers (in cm^{-1}) are indicated above each peak. The far infrared spectra of $BaCl_2 \cdot H_2O$ and $BaCl_2 \cdot 2H_2O$ are shown for comparison.

Table 9.5. Vibration Frequencies and Tentative Assignments

Salt	H$_2$O Librations[a]	M—O Stretch	H-Bond Stretch	Lattice Modes
CoCl$_2$·6H$_2$O	776—R, 656, 560, 464	341, 288[b]	216, 184, 136[c]	104, 80
NiCl$_2$·6H$_2$O	728—R(650—W),[d] 584, 480, 396[c]	(350),[d] 296[b]	200, 136[c]	80, 66, 56
CaCl$_2$·6H$_2$O	656, 480	380	240, 192	136, 100, 80
SrCl$_2$·6H$_2$O	656, 500	(350),[d] 312[b]	224, 160	112, 80
CrCl$_3$·6H$_2$O	~730—R, 590—W, 384—T	464	248, 160[e]	120, 72, 44, 32
AlCl$_3$·6H$_2$O	790—R, 570—W. 300—T[c]	465[c]	216,[c] 168[c]	128, 110, 65, 44
MgSO$_4$·7H$_2$O	665—R, 595—T, 485—W	380[c]	247, 165[e]	102, 70
MgCl$_2$·6H$_2$O	615, 536, 464	384[c]	200, 162[e]	112
FeCl$_2$·4H$_2$O	725, 552, 456	379, 320[c]	184, 141	76, 64
Al$_2$(SO$_4$)$_3$·18H$_2$O	810—R, 585—W, 320—T[c]	440[c]	220, 160[e]	<160
Al(NO$_3$)$_3$·9H$_2$O	790—R, 595—W, 310—T[c]	475[c]	250, 176	100
KAl(SO$_4$)$_2$·12H$_2$O	810—R, 585—W, 310—T, 710	480[c]	240, 196, 165[e]	107

[a] R = rocking; T = twisting; W = wagging.
[b] M—Cl stretching mode.
[c] See discussion in text.
[d] Very weak or unresolved transition.
[e] Possibly OMO deformation.

molecule (type II). The oxygen of the H_2O molecule is also in the sphere of coordination of the Ba^{++} ion. The neutron time-of-flight spectrum and the infrared spectrum of $BaCl_2 \cdot 2H_2O$ can be compared in Figure 9.13.[26] There is good agreement between the torsional frequencies observed in both measurements. Two additional peaks at 224 and 176 cm^{-1} are observed in the neutron spectrum. They are assigned to translational vibrations which involve the H_2O molecule and an oxygen or a chlorine ion, respectively. Both of these vibrations would require stretching of the hydrogen bond, and, if harmonic motions are assumed to be present, the ratio of the two frequencies is in keeping with such an assignment. The study of these low-frequency motions is important because they have been shown to be quite sensitive to the environment and degree of bonding of the H_2O molecule. The results shown in Table 9.5 indicate that the vibration frequencies of H_2O molecules are strongly dependent upon the degree and strength of the hydrogen bond with neighboring ions or H_2O molecules. Small, highly charged ions such as Al^{+++} may cause a change in the electrostatic equilibrium of the H_2O molecule, which results in an increase in the H—O—H angle and therefore in a shift in torsion frequencies. An attempt was made by Boutin et al.[26] to identify metal-oxygen (of H_2O) stretching frequencies. In the case of Al salts, a stretching frequency of about 440–480 cm^{-1} was consistently observed. This result will be used in a subsequent chapter to identify one of the sites on the surface of alumina powder.

9.5 Study of Diffusive Motions

It is difficult to distinguish clearly with techniques such as NMR, calorimetry, X-ray diffraction, or dielectric-constant measurements between a passage from an ordered state to a dynamically disordered phase (molecules reorienting almost freely) or to a statistically disordered phase (molecules in different but fixed orientations in different cells). In the case of neutron scattering, if the intermolecular forces hindering rotation in solids are very weak and a large fraction of molecules are able to surmount this potential barrier, the spectrum should show a very broad peak centered near the elastic peak, the width of which diminishes markedly with decreasing κ (momentum transfer). This broadening of the elastic peak has been studied in cyclohexane by Becka.[27] As the temperature was increased above the solid state transition ($186°K$), dynamic disorder was shown to set in; from the broadening of the elastic peak a coefficient of self-diffusion

D_s was calculated, where a "small motion" mechanism $D_s = 4.5 \times 10^{-5}$ cm^2/sec was assumed. From NMR data on spin lattice relaxation time, $D_s = 1.38 \times 10^{-5}$ cm^2/sec was obtained.[28] The discrepancy between the two values was ascribed to the effect of irreversible rotation of the molecules.

In addition to the determination of the diffusion coefficient, neutron studies of diffusion in H-bonded liquids have yielded other results of interest. If it is assumed that the diffusion coefficient D can be described in terms of an activation energy ϵ_0 such that $D = C \exp(-\epsilon_0/kT)$, values of the logarithm of the diffusion coefficient obtained from neutron measurements for water plotted versus $1/T$ give a straight line with $\epsilon_0 = 3$ kcal/mole, in good agreement with values of H-bond energies obtained by other methods.[29] A similar plot for glycerol as shown in Figure 9.14 reveals two markedly different slopes, with $\epsilon_0 = 7.5$ kcal/mole for $T > 100°C$ and $\epsilon_0 = 3$ kcal/mole for $T < 100°C$. The higher-temperature values for D are in good agreement with values obtained from the Eyring formula,[30] whereas at 20°C the neutron values are two orders of magnitude too large. The low-temperature behavior is attributed to the breaking of a single hydrogen bond, which leads to a partial rotation about a bond angle but with a lifetime much

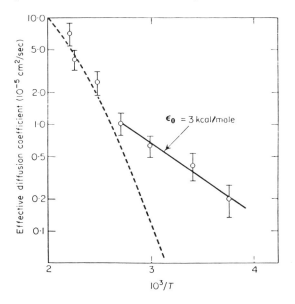

Figure 9.14. Effective diffusion coefficient versus $1/T$ for glycerol.

shorter than the delay time for diffusion; thus, simple diffusion is not the rate-determining process for proton motions. With increasing temperature a point is reached where three hydrogen bonds are broken so that the molecule can diffuse, and this becomes the rate-determining process. This approach should be equally valuable for obtaining information on molecular dynamics of adsorbed molecules.

Studies of diffusive motions in solids by means of neutron scattering have only recently been reported. Skold and Nelin[31] have studied the diffusion of hydrogen in palladium in the α phase and have compared their results with the theoretical prediction for jump diffusion.[32] From their measurements they concluded that the hydrogens were probably in octahedral sites and that the activation energy for diffusion was 3.7 ± 0.5 kcal/g-atom. This value is also obtained from NMR studies,[33] whereas values from macroscopic methods have ranged from 5 to 20 kcal/g-atom. Measurements with single crystals will eventually give direct information on the direction of proton jumps.

Infrared spectroscopy provides results somewhat similar to those from inelastic scattering of neutrons. In the long-wavelength region optical selection rules often drastically curtail the observation of some vibrations, although the resolution of the measurements is better at the present time than that of the neutron measurements. Infrared results in the study of surfaces have recently been reviewed by Burwell and Peri.[34] These studies have usually sought to identify the nature of the adsorbed species by observing the perturbation of vibrational molecular states when adsorbed. Several of these studies have been concerned with the adsorption of water vapor and the identification of hydroxyl groups on metal oxide adsorbents, and have involved observation of changes in the OH-stretching fundamentals and combination bands in the $1-3\,\mu$ region of the spectra. For example, if OH-stretching and H_2O fundamental vibrations can be differentiated, the presence of molecular water and different types of hydroxyl groups as a function of temperature and coverage can sometimes be inferred. The frequency shift due to hydrogen bonding in the OH fundamental is a measure of hydrogen-bond strength.[35] Investigation of the effect of adsorption on the rotational motion of molecules is also possible with infrared. The distortion at the surface can reduce the symmetry of the adsorbate molecule and result in the splitting of degenerate levels, allowing them to be observed. Neutron scattering has been used to identify the degree of rotational motion of adsorbed molecules (such as NH_3, CH_4, H_2O)[36] and to identify the state of adsorbed molecules (for example, H_2O

molecule, distorted H_2O, or hydrogen-bonded OH group).[37, 16] Optical spectroscopy provides only indirect information about bondings, derived from the values of the frequencies of fundamental vibrations of the H_2O molecules. The frequency of OH-stretching vibrations (v_1) has been shown to decrease in a regular manner with the strength of hydrogen bonding, as does the bending vibration (v_2), although to a lesser degree.[35] The fundamental frequencies do not characterize an H_2O molecule or an H-bonded OH group as unequivocally as do the frequencies below 800 cm^{-1}, which are accessible to the neutron technique. These correspond to torsional oscillations (800–400 cm^{-1})[16] or translations (approximately 200 cm^{-1}; for example, an OH—O stretching involving the hydrogen bond of the molecule as a whole. An OH group does not give rise to these same frequencies in the neutron spectrum. It has been shown in Chapter 7 how OH groups manifest themselves in the spectrum, as, for example, in the case of alkali hydroxides. Studies of the molecular dynamics of adsorbed molecules are presented in more detail in the next chapter.

REFERENCES

1. H. Hahn, in *Inelastic Scattering of Neutrons in Solids and Liquids* (International Atomic Energy Agency, Vienna, 1965), Vol. 2, p. 279.
2. H. Prask, H. Boutin, and S. Yip, *Proceedings of the Mid-American Symposium on Spectroscopy,* Chicago, May, 1967 (to be published); *J. Chem. Phys.* **48**, 3367 (1968).
3. E. Forslind, *Swed. Cement Concrete Inst., Roy. Inst. Technol., Stockholm, Proc.,* No. 21 (1954).
4. D. J. Hughes, H. Palevsky, W. Kley, and E. Tunkelo, *Phys. Rev.* **119**, 872 (1960). R. L. Stearns, H. R. Muether, and H. Palevsky, *Bull. Am. Phys. Soc.* **6**, 71 (1961). K. E. Larsson and U. Dahlberg, in *Inelastic Scattering of Neutrons in Solids and Liquids* (International Atomic Energy Agency, Vienna, 1963), Vol. 1, p. 317; *Reactor Sci. Tech.* **16**, 81 (1962).
5. Review by N. Ockman, *Advan. Phys.* **7**, 199 (1958).
6. W. Bagdade, *Proceedings of the Mid-American Symposium on Spectroscopy,* Chicago, May, 1967 (to be published); J. E. Bertie and E. Whalley, *J. Chem. Phys.* **46**, 1271 (1967).
7. E. Whalley and J. E. Bertie, *J. Chem. Phys.* **46**, 1264 (1967).
8. L. Slaggie, *Nucl. Sci. Eng.* **30**, 199 (1967).
9. M. S. Nelkin, *Phys. Rev.* **119**, 741 (1960).
10. T. Springer, *Nukleonik* **3**, 110 (1961); T. Springer and W. Wiedmann, *Z. Physik* **164**, 111 (1961).

11. S. W. Peterson and H. Levy, *Acta Cryst.* **10**, 70 (1957).
12. A. J. Leadbetter, *Proc. Roy. Soc. (London)* **287A**, 403 (1965).
13. A. Sjolander, *Arkiv. Fysik.* **14**, 315 (1958).
14. P. Flubacher, A. J. Leadbetter, and J. A. Morrison, *J. Chem. Phys.* **33**, 1751 (1960).
15. W. F. Giauque and J. W. Stout, *J. Am. Chem. Soc.* **58**, 1144 (1936).
16. H. Prask and H. Boutin, Picatinny Arsenal Internal Rpt. 1966, Picatinny Arsenal, Dover, New Jersey; *Bull. Am. Phys. Soc.* **10** (6), 687 (1965). H. Prask and H. Boutin, paper presented at 152*nd National A.C.S. Meeting,* New York, 1966.
17. A. Szkatula and A. Fuliuski, Report of Institute of Nuclear Physics, Krakow, December, 1966.
18. H. Stiller, in *Inelastic Scattering of Neutrons in Solids and Liquids* (International Atomic Energy Agency, Vienna, 1965), Vol. 2, p. 189.
19. G. S. Safford and A. W. Naumann, *Saline Water Conversion Report for* 1965 (U.S. Dept. of the Interior, Washington, D.C., 1966), p. 29.
20. G. Walrafen, *J. Chem. Phys.* **44**, 1546 (1966) and references therein.
21. F. E. Harris and C. T. O'Konski, *J. Phys. Chem.* **61**, 310 (1957).
22. J. C. Hindman, *J. Chem. Phys.* **36**, 1000 (1962).
23. H. S. Frank and W. Y. Wen, *Discussions Faraday Soc.* **24**, 133 (1957).
24. Th. G. Kujumzelis, *Z. Physik* **110**, 742 (1938).
25. M. S. Berqvist and E. Forslind, *Acta Chem. Scand.* **16**, 2069 (1962).
26. H. Boutin, G. J. Safford, and H. R. Danner, *J. Chem. Phys.* **40**, 2670 (1964); H. Prask and H. Boutin, *J. Chem. Phys.* **45**, 699 (1966); H. Prask and H. Boutin, *J. Chem. Phys.* **45**, 3284 (1966).
27. L. N. Becka, *J. Chem. Phys.* **38**, 1685 (1963).
28. D. W. McCall, D. C. Douglass, and E. W. Andrews, *J. Chem. Phys.* **31**, 1555 (1959).
29. L. Pauling, *The Nature of the Chemical Bond* (Cornell University Press, New York, 1960), 3rd ed.
30. S. Glasstone, K. J. Laidler, and H. Eyring, *The Theory of Rate Processes* (McGraw-Hill Book Company, Inc., New York, 1941).
31. K. Skold and G. Nelin, *Solid State Commun.* **4**, 303 (1966).
32. K. E. Larsson, in *Thermal Neutron Scattering*, edited by P. A. Egelstaff (Academic Press, Inc., New York, 1965), p. 347.
33. R. E. Norberg, *Phys. Rev.* **86**, 745 (1952).
34. R. L. Burwell, Jr. and J. B. Peri, *Ann. Rev. Phys. Chem.* **15**, 131 (1964).
35. G. C. Pimentel and A. L. McClellan, *The Hydrogen Bond* (W. H. Freeman and Company, San Francisco, 1960).
36. J. Janik and A. Kowalska, in *Thermal Neutron Scattering*, edited by P. A. Egelstaff (Academic Press, Inc., New York, 1965), p. 453.
37. H. Boutin and H. Prask, *Surface Sci.* **2**, 261 (1964).

10. Studies Related to the Adsorption of Hydrogenous Molecules on Surfaces

10.1 General

It is necessary to have a knowledge of the interatomic and inter-molecular forces between adsorbate and adsorbent and between adsorbate molecules in order to characterize adsorption processes. Similarly, bulk properties and forces between adsorbent molecules near the interface may have considerable influence on the properties exhibited by the surface. This is especially true in the case of semi-conducting elements and oxides. Hence, a knowledge of bulk properties such as conductivity, crystallinity, and particle size is important for determining surface behavior. The heterogeneity of the surface itself with respect to lattice vacancies, impurities, interstitial atoms, edges, corners, and dislocations can be induced or modified by temperature treatment or sample preparation. The identification of the "sorbed" species as being on the surface or in the bulk is, of course, essential and will be discussed in some detail in this section.

Depending on the nature of the interaction, two types of adsorption have in general been recognized.[1] In physical adsorption van der Waals' forces are involved. The heat of adsorption is small (0–10 kcal/mole) and multilayer coverage can occur. On the other hand, chemisorption involves bonding similar to that found in chemical compounds. The heat evolved is high ($\sim$ 15–25 kcal/mole) and the coverage is restricted to a monolayer. The forces at the surface also determine the mobility of adsorbed molecules. An adsorbed layer where the activation energy

171

required for migration from site to site is less than kT is considered a mobile layer. At each instant during the formation of such a layer the adsorbed molecules conform to the Boltzmann distribution. In contrast, an adsorbed layer is localized or immobile if the activation energy for migration is much higher than kT, and for the interaction times involved in neutron experiments the adsorbed particles do not have an equilibrium distribution. Intermediate cases where there is a time lag in attaining an equilibrium distribution are also recognized. The amount of heat evolved during adsorption and its variation with coverage (θ) provide information on the degree of mobility of the adsorbed molecule at the surface.[2]

The determination of thermodynamic parameters of adsorbed molecules (as a function of surface coverage and temperature) provides a test of physical models for the adsorbed state.[1] The parameters that can be derived from experimentally determined isotherms and isobars are the heats of adsorption, energies of activation, and free energies of the adsorbed molecules. The thermodynamic properties of a model are usually found from application of statistical mechanics through the use of the partition function Z_s, as shown in Chapter 9. Various assumptions as to the nature of the model and the sites on which adsorption occurs make it possible to evaluate the partition function and thereby the thermodynamic parameters.

Adsorption of hydrogenous molecules on a nonhydrogenous substrate is well suited to study by neutron scattering. In the case of adsorbed water molecules it is possible not only to study the degree of binding to the surface but also to determine in the case of multilayer adsorption the state of an assembly of H_2O molecules as compared, for example, to bulk water and ice. The state of the adsorbed water molecules may change, depending upon the number of layers adsorbed. For example, anomalies in the dielectric constant of water molecules adsorbed on the surface of LiF have been reported by Lajzerowicz et al.[3] and attributed to a phase change of the adsorbed water, conditioned essentially by the number of layers adsorbed.

In neutron-scattering studies the frequency distribution of the adsorbed molecules can be obtained at various stages of the adsorption. If only a monolayer is adsorbed at the surface, information about the different sites and the vibrational motions of H_2O molecules in those sites can be determined. At higher coverage one obtains only information about the average behavior of a large number of species.

Magnetic-resonance techniques are widely used in surface chemistry studies, particularly in determining the nature of the adsorbed species.[4] The NMR technique is suited only for studies on certain nuclei (for example, hydrogen), and ESR is limited to cases where a paramagnetic species is formed. Studies of surface water, hydroxyls, and other adsorbed hydrogenous and organic molecules have been made by NMR.[5] The narrowing of the linewidth of the proton-resonance spectrum indicates the existence or the onset of rotational motions, while spin-lattice relaxation times give reorientation rates.[6] In these measurements the variation of T_1, the longitudinal relaxation time, and T_2, the transverse relaxation time, with coverage and temperature reflects the influence of local fields upon the adsorbed species. If a given species can be differently adsorbed on a surface, each type of adsorption is characterized by a T_1 and T_2 value, and observation of a distribution of relaxation times indicates a multiphase adsorption. The bonding energy for the species might also be obtained from a study of the variations of T_2 with temperature. Zimmerman and co-workers[5] have investigated the nature of adsorbed water on the surface of high-area silica. Spin-echo techniques were employed by these authors to determine the longitudinal (T_1) and transverse (T_2) relaxation times, and this allowed a two-phase behavior of water to be established. The phase associated with a short relaxation time predominated at low coverages and hence had the largest heat of adsorption. Assuming the lifetime of a water molecule in an adsorbed phase to be the same as the lifetime of a hydrogen nuclear spin in a given relaxation phase, Zimmerman has calculated the lifetime of adsorbed water as $\tau = 3 \times 10^{-3}$ sec. Nuclear-magnetic-resonance studies of CH_4 adsorbed on TiO_2 have been made by Fuschillo and Aston.[7] An abrupt change in the proton-resonance width was observed at 20.4°K and for $\theta = 0.95$ and was attributed to translational and rotational diffusion of CH_4 molecules.

10.2   Water Adsorbed on Oxide Powders

10.2.1   *Introduction*

If one has an oxide with all its water adsorbed (at the surface) and none of it in the bulk, the adsorbed water can be ideally studied by neutron scattering. Unfortunately, this is not usually the case, since most oxides do contain water in the bulk. The extent of bulk water depends on the method of preparation and the initial thermal treatment. Complications also arise from the fact that any thermal treatment that reduces the bulk water also reduces the concentration of adsorbed water. However, for samples where the surface-to-volume ratio is rather high, that is, where the major part of the water present is at the surface, neutron-inelastic-scattering studies yield results pertinent to adsorbed water. Another useful technique is a preliminary outgassing at higher temperatures to remove most of the water in the sample (bulk and adsorbed), followed by readsorption of water at a lower temperature. In some cases the initial high-temperature treatment may reduce the surface area drastically because of sintering so that the concentration of water readsorbed at a lower temperature is not sufficient to give a good neutron spectrum. Perhaps the best method of studying surface water is to modify the preparation of the oxide itself. If 100 percent heavy water is used in preparing the oxide and if ordinary water is adsorbed following outgassing, the resulting system will have D_2O (if anything) in the bulk and H_2O at the surface. The neutron spectrum from such a system can then be solely attributed to the adsorbed H_2O since the scattering from D_2O molecules is much less.

The contribution to the scattered-neutron spectrum from the bulk material (adsorbent substrate) must also be taken into account. This can be minimized by using samples of very high surface area. A rough calculation in which the adsorbing particles are assumed spherical shows, however, that a monolayer of H_2O adsorbed on "Hisil" silica (surface area 160 m^2/g) gives only twice the scattering of the bulk material. For low-area rutile (6.4 m^2/g), the bulk gives approximately ten times more scattering than a monolayer of adsorbed H_2O. These values show that a careful characterization of the adsorbate species requires the precise determination of the bulk contribution with no adsorbate.

10.2.2 *Silica*

Neutron time-of-flight spectra obtained at room temperature for an unactivated sample (I) of silica ("Hisil," 160 m^2/g) and for a similar sample (II) which has been dried at 150°C for 12 hours are shown in Figures 10.1.[8] The time-of-flight spectrum of liquid water is shown for comparison in Figure 10.2. The contribution from species adsorbed in the bulk has not been subtracted. Substantial differences are observed between the neutron spectra of silica I and II even with this relatively mild activation.

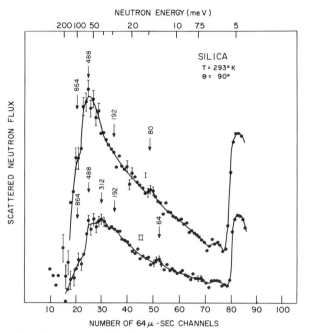

Figure 10.1. Neutron time-of-flight spectra of water adsorbed on silica. (a) Sample after exposure to air for several days. (b) Same sample after evacuating at 150°C for 12 hours. The indicated vibrational frequencies are in cm^{-1}.

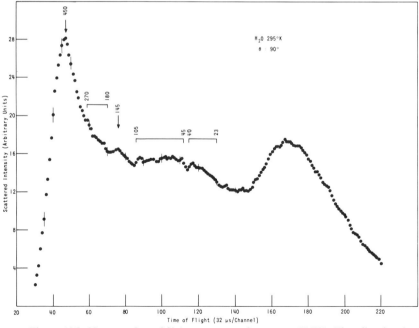

Figure 10.2. Neutron time-of-flight spectrum of water at 295 K. The vibrational frequencies are in cm^{-1}.

175

In the spectrum of silica I, torsional vibrations (488 cm^{-1}) and translational vibrations (the broad band extending from 192 to 80 cm^{-1}) of the adsorbed H_2O molecules are quite analogous to those of liquid water with, however, an extra peak at 864 cm^{-1}. The elastic peak (channel 80) is sharper than that of liquid water, and this suggests less mobility for the adsorbed water. The neutron spectrum of silica II, on the other hand, does not display any pronounced torsional peak in the region of 400 cm^{-1}, unlike other cases of bound water previously discussed, such as ice or crystal hydrates. The neutron spectrum of water adsorbed at the surface of silica after heating at 150°C for 12 hours is similar to that of water adsorbed in the zeolite cages of analcite.[9] The H_2O molecules in analcite have low mobility, and their coordination changes rapidly as a result of the mobility of the Na^+ ions in the cages. This suggests the existence of several sites on the surface of silica II. Two possible configurations for the adsorbed water which are consistent with the neutron data are shown in Figures 10.3a and b. A situation such as the one illustrated in Figure 10.3a with additional

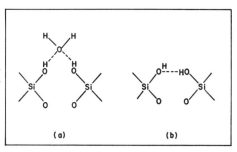

Figure 10.3. Possible configurations for the water molecule adsorbed on silica.

(a)   (b)

layers of water molecules or a variety of Si–Si distances would give rise to a spectrum similar to that of silica I. Upon outgassing or mild heating, a situation such as the one illustrated in Figure 10.3b would occur, giving rise to a spectrum like that of silica II. Benesi and Jones observed peaks at 810 and 870 cm^{-1} in the infrared spectrum of SiO_2 and adsorbed water, which they assigned to Si—O stretching and Si—OH bending vibrations.[10] This is in agreement with the broad band observed at 864 cm^{-1} in the neutron spectra of both silica I and silica II. Changes with temperature in the overtone combination bands and the OH-stretching fundamentals were studied by McDonald[11] and by Anderson and Wickersheim[12] for water adsorbed on silica gel. The latter identify four adsorbed species: monomeric H_2O, hydrogen-bonded H_2O, unbound OH, and hydrogen bonded on the surface of silica gel. Only OH groups seem to remain after heating at 150°C. This is in reasonable agreement with the interpretation of the neutron results, although it is difficult to distinguish between hydrogen-bonded OH groups and a distorted hydrogen-bonded H_2O molecule.

10.2.3 *Alumina*

Neutron-scattering studies of water adsorbed on γ alumina (Linde B, 0.005-μ particle size) have also been performed on a sample (I) exposed to air for several days (Figure 10.4) and on a similar sample (II) heated

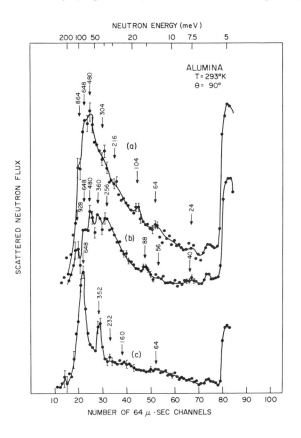

Figure 10.4. Neutron time-of-flight spectra of (a) γ alumina after exposure to air for several days, (b) γ alumina after evacuation for 12 hours at 150°C, (c) $AlCl_3 \cdot 6H_2O$. Vibrational frequencies are in cm^{-1}.

at 150°C for 12 hours.[8] Changes qualitatively similar to those of silica were observed upon activation. The neutron spectrum of alumina I (Figure 10.4a) is very similar in shape to that of liquid water. The peaks at 648 and 360 cm^{-1} in the spectrum of alumina II (Figure 10.4b) are assigned, respectively, to torsion and translation vibrations of an H_2O molecule coordinated to a surface Al^{+++} ion, since similar peaks are observed in the neutron spectrum of $AlCl_3 \cdot 6H_2O$ (Figure 10.4c). The similarity to the hydrate could be attributed to coordinated bulk water; however, thermogravimetric measurements seem to indicate that little bulk water is present in γ alumina.[13, 14] The sharpness of these peaks in the spectrum of $AlCl_3 \cdot 6H_2O$ is believed to result from the existence of only one type of H_2O in the hydrate, with the high frequency due to the very short $Al—OH_2$ distances (1.88Å). The peak at 480 cm^{-1} in the spectrum of alumina II indicates that some physisorbed water molecules remain at 150°C; the sharp peak at 928 cm^{-1}, which has no counterpart in the $AlCl_3 \cdot 6H_2O$ spectrum, is assigned, as in the case of silica, to a metal-OH bending vibration.

It is interesting to note that if the assignments of $Si—OH$ (864 cm^{-1}) and $Al—OH$ (928 cm^{-1}) bending are correct, a lower limit on the fraction of H_2O present as hydroxyl can be determined. The integrated intensities, after correction for thermal population, indicate that of the H_2O molecules present one-third or more appear to occur as OH on the activated silica and one-half or more as OH on the activated alumina.

Peri and Hannon have shown by infrared studies the existence of molecular water on the surface of γ alumina up to about 400°C.[15] De Boer *et al.* have studied the relation between lauric acid adsorption and specific surface area of rehydrated samples of γ alumina.[13] Under their conditions it appears that "chemisorbed" water molecules do not react with oxygen atoms in the surface-forming OH groups but rather are bound to surface oxygens by very strong hydrogen bonds. Maciver *et al.* studied both η and γ alumina by means of gas adsorption and immersion calorimetry.[14] They found that for γ alumina the observed behavior was consistent with adsorbed water being present entirely as hydroxyl groups. For γ alumina, on the other hand, at $T \leqslant 300°C$ an excess of water was found which was attributed to molecular water. This is in agreement with the infrared results of Peri and Hannon. The neutron spectral data suggest that at 120°C several species are present on the surface of alumina. These species could be represented as shown in Figure 10.5.

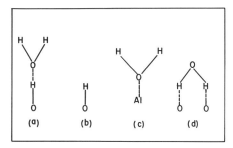

Figure 10.5. Possible arrangements of water molecules present on the surface of alumina.

10.2.4 *Rutile*

Neutron spectra have been obtained for a high-area rutile sample (160 m^2/g) exposed to air for several days and for the same sample after outgassing at 200°C for 24 hours.[16] The spectra are shown in Figures 10.6a and b, respectively. The ratio of the contribution of the bulk and a monolayer of adsorbed water to the scattering cross section

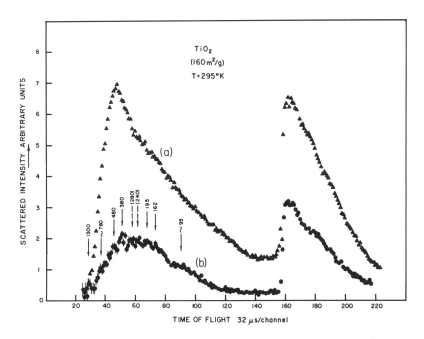

Figure 10.6. Neutron time-of-flight spectrum of high-area rutile (160 cm^2/g) (a) exposed to air for several days, (b) after outgassing at 200°C for 12 hours. Vibrational frequency in cm^{-1}.

for this high-area sample is approximately 0.4. The major contribution to the observed spectrum therefore comes from the H_2O molecules adsorbed on the surface, usually in more than a monolayer. Indeed, the unactivated sample of rutile indicates the presence of physisorbed molecular water on the surface, with a lower mobility than in pure water.

Infrared measurements as a function of temperature by Lewis and Parfitt[17] and by Yates[18] show the disappearance after outgassing at 200°C of a 1600 cm^{-1} band assigned to the fundamental deformation frequency v_2 of H_2O molecules. Thermogravimetric measurements also indicate a loss between 100°C and 400°C of strongly physisorbed molecular water along with OH-group condensation.[17] The neutron spectrum of the activated sample (Figure 10.6b) displays a peak at 480 cm^{-1}, indicating the presence of molecular water even after outgassing at 200°C. Neutron spectra of a very low-area (less than 0.5 m^2/g) rutile sample are shown in Figure 10.7. Figure 10.7a corresponds to the rutile sample after sintering at 1200°C; Figure 10.7b is the spectrum of the same sample after reduction in hydrogen at 500°C for two hours. For this very low-area sample, the expected scattering from the bulk material is more than two orders of magnitude greater than that from a monolayer of adsorbed H_2O. These spectra, therefore, correspond to the lattice modes of TiO_2 and are in excellent agreement with the infrared results of Liebisch and Rubens[19] and Waldron (as reported by von Hippel *et al.*[20]) as well as with the Raman measurements of Narayanan.[21] Waldron's infrared results show broad maxima at 600 and 260 cm^{-1} and a sharp peak at 410 cm^{-1}, whereas the neutron spectra shows peaks at 790, 570, 405, 290, 195, and 90 cm^{-1}. The differences observed in the neutron spectrum of the reduced sample are attributed to vibrations involving hydrogen already present in the bulk, enhanced by the additional adsorbed hydrogen. Von Hippel *et al.*[20] concluded from infrared studies of the OD- and OH-stretching region that hydrogen was present in the bulk of TiO_2. Two bands each are observed for OD-stretching and OH-stretching vibrations; these are assigned to deuterons or protons trapped by O^{--} and forming hydrogen bonds at two different sites. Upon hydrogen reduction the transitions at 790 and 195 cm^{-1} approximately double in intensity, the latter becoming very broad. These transitions are assigned as corresponding, respectively, to TiOH bending (in analogy with silica) and OH---O stretching vibrations.

The absence of peaks at 405 and 290 cm^{-1} in the activated high-area

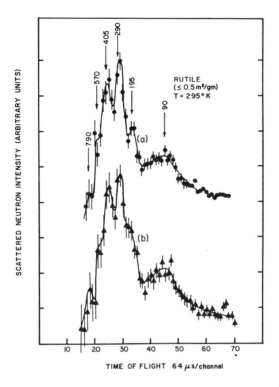

Figure 10.7. Neutron time-of-flight spectrum of low area rutile (0.5 cm^2/g) after (a) sintering at 1200°C and after (b) reduction in hydrogen at 500°C for 2 hours. Vibrational frequency in cm^{-1}.

TiO$_2$ spectrum confirms that the spectrum is due primarily to surface H$_2$O. The shoulder indicated at 780 cm^{-1} is attributed to TiOH bending vibrations, and its broadness indicates a multiplicity of surface OH species. The weak transitions at 195 and 162 cm^{-1} are probably due to OH−−−O stretching vibrations of surface species and might arise, for example, from linkages of the types shown in Figures 10.5a and d.

The examples considered thus far in this chapter illustrate the type of information that can be provided by neutron studies when groups such as hydroxyl (OH) are present in the bulk of a material or adsorbed on the surface. It seems possible to differentiate between OH groups in the bulk, which in most cases reflect (and amplify) the vibrations of the lattice with no change upon deuteration, and H$_2$O molecules in the bulk, which can be identified by the shift of the characteristic torsion frequency upon deuteration. Hydroxyl groups, on the other hand, are either on the sphere of coordination of a metal or hydrogen-bonded to an anion at the surface. The stretching frequencies involved in the OH motion should not be affected by deuteration.

10.3   Diffusive Motions

Neutron-inelastic-scattering studies of diffusive motions have also been performed with methane and hydrogen adsorbed on high-surface-area charcoal (110 m^2/g).[22] These measurements show that for adsorbate concentrations of $\frac{1}{7}$ to $\frac{1}{8}$ monolayer capacity the inelastic portion of the spectra of adsorbed molecules is very similar to the spectra of the pure liquids. The results for methane are shown in Figure 10.8. This spectrum represents the difference between the total spectrum and the spectrum of charcoal alone. The results of a cross-section calculation,

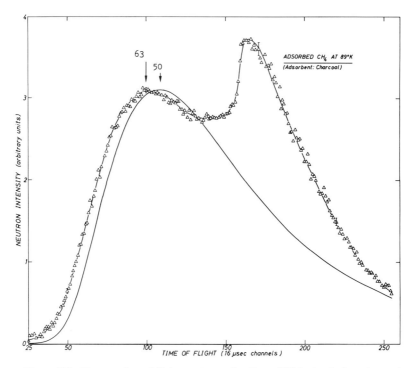

Figure 10.8. Neutron time-of-flight spectrum of methane (CH$_4$) adsorbed on charcoal at 89°K. The solid line is a calculated spectrum assuming a free gas model. The frequencies shown above the inelastic peak are in cm^{-1}.

in which a free gas model is assumed, are also shown for comparison. Hydrogen adsorbed on the same substrate shows very similar behavior. The deviation from the Kreiger-Nelkin model[23] is attributed in the case of methane to possible Debyelike quasi-crystalline behavior which could cause a shift in the spectrum. An alternate explanation might be that the CH_4 translation vibration that is expected to occur at about $100-200$ cm^{-1} is broad because of anharmonicity and surface heterogeneity, and this broadening causes an apparent shift. The translation vibration of H_2 against the surface is expected to occur at 320 cm^{-1} but is not observed, probably because of poor thermal population.

Both adsorbates showed sharp quasi-elastic peaks with only very slight broadening, an effect suggesting that diffusion on the surface takes place after a long residence time. This finding contrasts with the spectra observed for pure liquid methane in which at the same 90° scattering angle no quasi-elastic peak is observed. The absence of a peak indicates that the diffusion constant is much larger.[22]

The studies cited here of hydrogen and methane adsorbed on charcoal were made at a single scattering angle so that a detailed quantitative determination of jump times and diffusion constants was not possible.

10.4 Role of Water in Biological Systems

There is much uncertainty about the state of H_2O molecules in the vicinity of a protein molecule and about the role played by H_2O molecules in various biological processes. Studies of systems of biological significance by Raman scattering,[24] nuclear magnetic resonance,[24] and dielectric measurements[25] seem to indicate a decrease in the rate of motion of the water protons relative to their rate in pure water and an increase in the degree of hydrogen bonding of the H_2O molecules.

A determination of the energy involved in bonding H_2O molecules to the protein molecule is difficult because it cannot be clearly separated from the interactions taking place between polar or nonpolar groups within the protein molecule or between adjacent protein molecules (hydrogen bonds, hydrophobic bonds, van der Waals' forces). Attempts to estimate these inter- and intra-molecular interactions have been made based on studies of the transition between two conformations of the polypeptide molecule (that is, α helix/random coil transition).[26] Unfortunately, these studies must be performed in solution and are affected by all the uncertainties mentioned earlier. A neutron-scattering study of such transitions made on polypeptides in the solid state[27] has been described in Chapter 5.

Figure 10.9 was obtained by subtracting the spectrum of the sodium

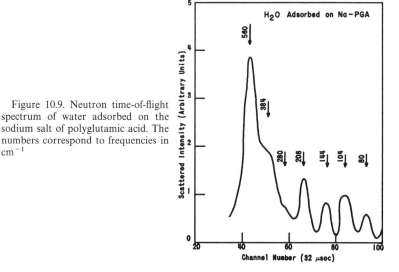

Figure 10.9. Neutron time-of-flight spectrum of water adsorbed on the sodium salt of polyglutamic acid. The numbers correspond to frequencies in cm^{-1}

salt of polyglutamic acid (random coil) exposed to water vapor at atmospheric pressure from its spectrum after dehydration. This difference spectrum corresponds to water adsorbed, probably as a monolayer, at the peptide group or near the polar end of the side chain. It seems to indicate that the H_2O molecules are relatively strongly bound, with a torsional peak at 560 cm^{-1} and three peaks between 100 and 200 cm^{-1}. These peaks might be assigned (by analogy with H_2O molecules in hydrates or ice) to translational vibration of the adsorbed H_2O molecules, probably at different sites on the polypeptide molecules.

Finally, some characteristic vibrations of a polypeptide chain (such as amide VII torsion) have been observed by neutron scattering and have been shown to be quite frequency-sensitive to the selective adsorption of H_2O molecules.[28]

10.5   Study of Hydrogenous Molecules in Zeolite and Molecular Sieves

The study of molecules such as H_2O adsorbed in the molecular cages of natural zeolites lends itself to neutron-scattering techniques.[9] In natrolite ($Na_2Al_2Si_3O_{10} \cdot 2H_2O$) there is on the average one H_2O molecule per cavity, and it is hydrogen-bonded to oxygens of the framework. The neutron spectrum (Figure 10.10a) displays very strong

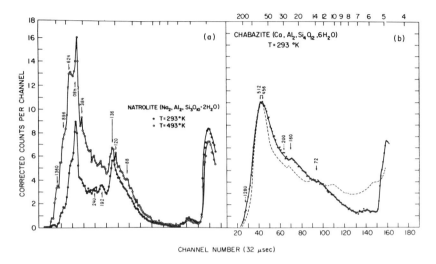

Figure 10.10. Neutron time-of-flight spectrum of natrolite and chabazite. The dashed curve in the chabazite spectrum is that of liquid water. The vibrational frequencies are in cm^{-1}.

peaks corresponding to libration and translation frequencies, and this suggests a strongly bonded H_2O molecule vibrating as in a lattice. In chabazite ($CaAl_2Si_4O_{12} \cdot 6H_2O$), on the other hand, where on the average there are 20 H_2O molecules per cage, the neutron spectrum is broader (Figure 10.10b) and quite similar to that of liquid water. This finding is interesting because in adsorption studies a chemisorbed H_2O molecule is expected to behave like one in natrolite, whereas the physisorbed H_2O molecules display a spectrum more like that of water. The spectrum of a sample of beryl containing 1 percent adsorbed water is also of interest because at room temperature it is very broad and suggests a large degree of rotational freedom among the H_2O molecules (Figure 10.11). A measurement of the neutron spectrum at low temperatures, however, indicates that at least some of the H_2O molecules actually undergo a weakly hindered oscillation with a potential barrier much lower than in the previous cases, and no hydrogen bonds formed with neighboring molecules.[29]

Very recently Egelstaff and co-workers have performed neutron-

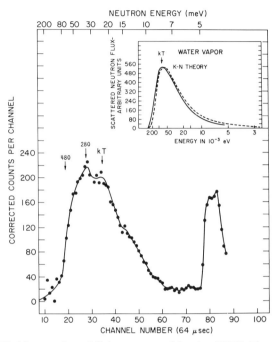

Figure 10.11. Neutron time-of-flight spectrum of Beryl at 293°K. The neutron spectrum of water vapor is shown in the insert for comparison. The peak frequencies are indicated in cm^{-1}.

scattering studies on H_2O, D_2O, methyl alcohol, methylcyanide, and ammonia adsorbed in Linde Molecular Sieve 3A.[30, 31] A theoretical development was presented in which an adsorbed molecule was described by two independent frequency distributions corresponding to vibratory motions parallel and perpendicular to the adsorbing substrate. A careful analysis of diffusive motions was made, and because the width of the quasi-elastic peak was found to be almost constant with respect to changes in the momentum transfer (κ), a specific model for jump diffusion was used in interpreting the results. After fitting the experimental curves, jump times and diffusion constants were obtained for CH_3CN, CH_3OH, and NH_4^+, as shown in Table 10.1.[31] The diffusion constant for liquid methylcyanide was also obtained and found to be 4.7 ± 10^{-5} cm^2 sec^{-1}. Table 10.1 shows the jump time to be about 10^{-11} sec. This value contrasts with a value of 10^{-6} sec for the nuclear relaxation time of water in 4A and 13X molecular sieves from NMR measurements, the latter time being associated with jumps

Table 10.1. Diffusion Constants and Jump Times of Molecules Trapped
in Zeolite Cavities

Trapped Molecule	Diffusion Constant	Jump Time
CH_3CN	$< 1.5 \times 10^{-5}$ cm^2 sec^{-1}	$1.5 \pm 0.5 \times 10^{-11}$ sec
CH_3OH	$1.3 \pm 0.5 \times 10^{-5}$ cm^2 sec^{-1}	$1.0 \pm 0.5 \times 10^{-11}$ sec
NH_4^+	$< 1.5 \times 10^{-5}$ cm^2 sec^{-1}	$1.3 \pm 0.4 \times 10^{-11}$ sec

of molecules from one lattice site to another.[32] A correlation time of 10^{-8} sec was also found from NMR measurements and ascribed to molecular rotations of molecules not attached to cage walls. The neutron results indicate the presence of another jump process, possibly due to the dissociation-reformation of a hydrogen bond. Similar types of proton motions have been observed by Stiller for a solution of HF in water.[33]

These examples illustrate the sensitivity of neutron scattering to the details of molecular rotational motion and to the degree of binding of molecules. The possibility of distinguishing between free and bound water, for example, and the ability to determine the potential barrier of the various states is of great importance to a large number of problems in surface and colloid chemistry.

10.6   Study of Clathrate Hydrates

When water is crystallized slowly in the presence of certain gases or solutes, a modified form of ice results, with cavities that usually contain one molecule each.[34, 35] These typically nonpolar guest molecules stabilize the cavities. The oxygen lattice in a clathrate hydrate is symmetrically distinct from any of the known crystalline modifications of pure ice, and some of those hydrates have been studied by X-ray diffraction.

Apart from the X-ray structural work, most information about clathrate hydrates has come from phase stability studies—chemical analysis and dissociation pressures. Thus the thermodynamics of these materials is reasonably well known, as are the heavy atom positions, and there are some theoretical treatments of the requirements for cavity stability.[36] Beyond this, the complete structural description of the clathrate hydrates and their relation to ordinary ice, along with the picture this information can give of the binding properties of water molecules in various condensed phases, makes it important to understand what intermolecular forces are responsible for the stability of these hydrates and to what extent the hydrogen atoms and hydrogen bonds contribute to their structure and stability.

Several important questions must be answered before this understanding can be reached:[37]

1. What are the vibrational modes and frequencies of the lattice and how do these differ among clathrate hydrates? How do they differ from the corresponding motions of ordinary ice and of the water sublattices in salt hydrates?

2. What is the rotational state of the guest molecules at different temperatures, and what is their degree of translational freedom in the cavity?

3. What is the nature and strength of the interaction between guest and lattice wall which stabilizes the cavity structure? How are the states of guest and lattice molecules modified upon crystallization of a hydrate, and what effect (including the possibility of partial disorder) does this have on the thermodynamic properties, especially the heat capacity, of the system?

As was shown by the examples given in the previous sections, the neutron-scattering technique seems well suited to provide answers to these questions. A nonpolar molecule in the water phase produces a large local change in the internal pressure, $(dU/dV)_T$. Clathrate hydrates can thus be used as models for the stability of different packings of water molecules, especially when coupling force constants are derived from spectroscopic analysis.

With respect to the extent of the ice lattice, the various materials can be thought of as constituting a series ranging from pure ice to the nearly independent water molecule in salt hydrates. Some hydrates consist of a true, if modified, ice lattice in which nonbinding repulsive forces between the guest and water molecules stabilize the cavities by essentially making the lattice wall energetically more favorable than a structure with intimate guest-water contacts. With a more polar guest such as an amine a different lattice is stabilized and, if geometry is favorable (as with hexamine), some guest-lattice hydrogen bonding appears. The effect of this rearrangement on the hydrogen motions is not known in any detail. Introduction of the proper ions in some cases can stabilize cavities if the counter-ion can be incorporated in the ice lattice, a condition that is met by hydroxide and fluoride. The next step completely breaks down the ice lattice, so that a hydrate such as $MgCl_2 \cdot 12H_2O$ is best regarded as a packing of hydrated ions, each surrounded by six octahedrally coordinated water molecules; and, of course, water molecules in neighboring coordination shells interact. Finally, the classical salt hydrates can be regarded as ionic lattices with water molecules included.

Throughout such a series, vibration frequencies vary with packing distance and orientation and with the corresponding degree of coupling. When these frequencies and the corresponding band shapes are interpreted according to a structural model for the crystal, proton structures can be determined and interaction potentials in the several structures evaluated. The potential enters into the partition function, so that it is relevant to predicting all the equilibrium properties of the structures, including the range of stability.

The hindered rotational and translational states of the guest molecules also provide a test of some of the theories of clathrate stability and are important in estimating the selectivity of the lattice toward guest molecules—in other words, the purity, so relevant in water purification schemes.[35] The use of guest molecules in clathrate hydrates as models of nonpolar residues of bipolymers in aqueous solution also bears on the state of the guest, for here it is the organic residue that has biological activity, no matter what the role of water in enhancing or destroying this activity.[38]

Many small molecules trapped in inert-gas matrices rotate freely, even at cryogenic temperatures.[39] The ammonium ions in crystals of their halide salts also rotate freely above definite phase-transition temperatures.[40] Molecular geometry seems to be related to both dielectric relaxation and crystal plasticity.[41] In some theoretical calculations[36] small, nearly spherical molecules in some hydrates are assumed to rotate freely, for this leads to better-calculated dissociation pressures; however, the degree of rotational freedom has never been directly observed. Neutron scattering is quite sensitive to rotational freedom and offers a convenient tool in this area (see Chapter 8). Hexamine, for example, is a roughly spherical molecule, but it is connected to its hydrate lattice by three hydrogen bonds. The potential barrier to rotation, the possibility of cooperative phenomena between guest molecules, and the interaction between water and guest molecules must be considered if the complete dynamics of the lattice is to be understood. Thus the correlation of rotational freedom with degree of polarity of the guest molecule for the series methane, methylamine, tetramethylammonium hydroxide hydrates could make it possible to derive a general potential form and to predict the freedom in other hydrates.

The hindered translational motions of a guest molecule in a cavity is related to the case of diffusional jumps, and the mode of the vibrations can be used to test the theoretical prediction that small guests are isolated in the center of the cavity and only rarely collide with the lattice wall.

Finally, the breadth of the vibrational bands of the lattice is related to details of coupling in the system and to disorder in the packing. The protons in ordinary ice are disordered, and this produces a distortion in the oxygen lattice because of local asymmetries in the binding forces.[42] The protons in methane hydrate are also disordered, but the structure is not well enough known for the effect on the lattice of this ice to be estimated.

The low-frequency spectra of the clathrate hydrates can be related to the structures of the pure ices, to those of salt hydrates, and to aqueous solutions. It should be noted again that interpretation of neutron-scattering spectra is greatly facilitated if complementary infrared and Raman spectra are also available;[43] neutron and optical techniques in conjunction can specify both vibrational modes and potentials.

Clathrate hydrates are of increasing importance and offer a fascinating and largely untouched field of research, especially with respect to dynamics of motion of the guest molecules.

REFERENCES

1. See, for example, A. W. Adamson, *Physical Chemistry of Surfaces* (Interscience Publishers, New York, 1960).
2. S. Brunauer, P. H. Emmett, and E. Teller, *J. Am. Chem. Soc.* **60**, 309 (1938).
3. J. Lajzerowicz, P. Ducros, and J. Deville-Cavillin, *Phys. Letters* **3**, 248 (1963); J. Lajzerowicz, *Compt. Rend.* **253**, 234 (1961).
4. D. E. O'Reilly, *Advan. Catalysis* **12**, 31 (1960).
5. J. R. Zimmerman, B. Holmes, and J. Lasater, *J. Phys. Chem.* **60**, 1157 (1956); J. R. Zimmerman and W. E. Britt, *J. Phys. Chem.* **61**, 1328 (1957); J. R. Zimmerman and J. A. Lasater, *J. Phys. Chem.* **62**, 1157 (1958).
6. N. Bloembergen, E. M. Purcell, and R. V. Pound, *Phys. Rev.* **64**, 680 (1946).

7. N. Fuschillo and J. G. Aston, *J. Chem. Phys.* **24**, 1277 (1956); N. Fuschillo and J. G. Aston, *Nature* 1277 (1956).
8. H. Boutin and H. Prask, *Surface Sci.* **2**, 261 (1964).
9. H. Boutin, G. J. Safford, and H. R. Danner, *J. Chem. Phys.* **40**, 2670 (1964).
10. H. A. Benesi and A. C. Jones, *J. Phys. Chem.* **63**, 179 (1959).
11. R. S. McDonald, *J. Am. Chem. Soc.* **79**, 850 (1957).
12. J. H. Anderson and K. A. Wickersheim, *Solid Surfaces,* edited by H. C. Gatos (North Holland Publishing Company, Amsterdam, 1964), p. 252.
13. J. de Boer, J. M. G. Fortuin, B. C. Lippens, and W. H. Meijs, *J. Catalysis* **2**, 1 (1963).
14. D. S. Maciver, H. H. Tobin, and R. T. Barth, *J. Catalysis* **2**, 485 (1963).
15. J. B. Peri and R. B. Hannon, *J. Phys. Chem.* **64**, 1526 (1960).
16. H. Boutin, H. Prask, and R. D. Iyengar, *Advan. Colloid Interface Sci.* **2**, 1 (1968).
17. K. E. Lewis and G. D. Parfitt, *Trans. Faraday Soc.* **62**, 204 (1966).
18. D. J. C. Yates, *J. Phys. Chem.* **65**, 746 (1961).
19. T. Liebisch and H. Rubens, *Preuss. Akad. Wiss. (Berlin)* **8**, 211 (1921).
20. A. von Hippel, J. Kalnajs, and W. B. Westphal, *J. Phys. Chem. Soc.,* **23**, 779 (1962).
21. P. S. Narayanan, *Proc. Indian Acad. Sci.* **32A**, 279 (1950).
22. G. Verdan, private communication.
23. T. J. Krieger and M. S. Nelkin, *Phys. Rev.* **106**, 290 (1957).
24. J. Clifford, B. A. Pethica, and W. A. Senior, *Ann. N.Y. Acad. Sciences* **125**, 458 (1965).
25. E. H. Grant, *Ann. N.Y. Acad. Sciences* **125**, 418 (1965).
26. J. A. Schellman, *Compt. Rend. Trav. Lab. Carlsberg, Ser. Chim.* **29**, 233 (1955).
27. H. Boutin and W. Whittemore, *J. Chem. Phys.* **44**, 3127 (1966).
28. H. Boutin and V. D. Gupta, private communication.
29. H. Boutin, H. Prask, and G. J. Safford, *J. Chem. Phys.* **42**, 1469 (1965).
30. J. S. Downes, J. W. White, P. A. Egelstaff, and V. S. Rainey, *Phys. Rev. Letters* **17**, 533 (1966).
31. P. A. Egelstaff, J. S. Downes, and J. W. White, private communication.
32. J. P. Cohen Addad, *Brit. Radiospec. Group Conf.,* Canterbury, England (1966).
33. H. Stiller, in *Inelastic Scattering of Neutrons in Solids and Liquids* (International Atomic Energy Agency, Vienna, 1965), Vol. 2, p. 189.
34. J. H. Van der Waals and J. C. Platteuw, *Advan. Chem. Phys.* **2**, 1 (1959).
35. G. A. Jeffrey, *Saline Water Conversion Report for* 1965 (U.S. Dept. of the Interior, Washington, D.C., 1966), p. 86.
36. V. McKoy and O. Sinanoglu, *J. Chem. Phys.* **38**, 2946 (1963).
37. T. Wall, private communication.
38. W. Kauzmann, *Advan. Protein Chem.* **14**, 1 (1959).
39. M. T. Bowers and W. H. Flygare, *J. Chem. Phys.* **44**, 1389 (1966).
40. W. Vedder and D. F. Hornig, *J. Chem. Phys.* **35**, 1560 (1961).
41. C. P. Smyth, *J. Phys. Chem. Solids* **18**, 40 (1961).
42. J. E. Bertie and E. Whalley, *J. Chem. Phys.* **40**, 1637 (1964).
43. T. T. Wall and D. F. Hornig, *J. Chem. Phys.* **43**, 2079 (1965).

11. Molecular Systems of Special Interest

11.1 Acetamide and Related Compounds

The neutron spectrum of acetamide, shown in Figure 11.1a, displays a broad band centered at about 180 cm^{-1}, which is assigned[1] to the hindered rotation of the CH_3 group, and a shoulder at about 600 cm^{-1} (probably due to NH torsional frequency). Because of the large degree of rotational freedom of the CH_3 group, there are many transitions between thermally populated rotational levels with which the neutron can exchange energy. The most probable transition $(0 \rightarrow 1)$ for the torsional motions of the CH_3 group corresponds to the band centered at 180 cm^{-1}; if a threefold cosine barrier is assumed, the barrier-to-rotation height is estimated at about 2 kcal/mole. The intense band due to CH_3 rotation obscures any structure in the spectrum contributed by other modes of vibration. The neutron spectrum of trifluoroacetamide has also been obtained (Figure 11.1b), and with the CH_3 torsional band absent, other vibration frequencies can now be observed. Table 11.1 shows the observed infrared and neutron frequencies of trifluoroacetamide.[1] In this particular case neutron scattering cannot compete in resolution with infrared absorption, and all the frequencies cannot be observed, although there is reasonable agreement between the two sets of data. For example, the neutrons see only an intense broad band centered at 99 cm^{-1} in the low-frequency region of the spectrum which contains all the external optical modes, while the infrared data resolve all four of the active optical branches.

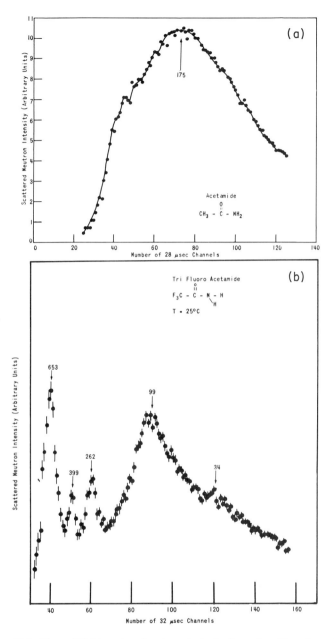

Figure 11.1. Time-of-flight spectra of neutrons inelastically scattered by polycrystalline samples of (a) acetamide and (b) trifluoroacetamide at 298°K and at 90° scattering angle. The energy transfers are indicated above each peak in cm^{-1}. The ordinates are proportional to the differential scattering cross section.

193

Table 11.1. Vibration Frequencies (cm^{-1}) in
Trifluoroacetamide $(293°K)$

Infrared Data	Neutron Data
90	34
115	99
130	140
153	
263	262
287	
400	400
440	
515	
600	
680	653
700	

The peak at 34 cm^{-1} in the neutron spectrum, however, might corre-spond to one or several acoustic branches that are not observed in infrared. Two peaks at 440 and 400 cm^{-1} in the infrared spectrum are not resolved in the neutron spectrum, where a single peak at about 400 cm^{-1} is observed.

Neutron data, despite insufficient resolution, can provide qualitative characteristic information other than the observation of acoustic branches of normal modes. For example, a comparison of the neutron spectra of acetamide and trifluoroacetamide indicates that (a) the torsional frequency of the CH_3 group occurs at 180 cm^{-1}; (b) the sharp peak at 262 cm^{-1} can be assigned to an amide VII band (because of its sharpness and intensity), in agreement with the results obtained for N-methylacetamide; and (c) because of its intensity the band at 653 cm^{-1} in trifluoroacetamide is due to NH_2 torsional vibration, a datum that could not be inferred from infrared measurements alone.

In N-methylacetamide (NMA), there are two methyl groups whose torsional frequencies are expected to be close. An intense and broad peak is observed in the neutron spectrum[1] centered at about 130 cm^{-1}, and this band is assigned to these torsions. A normal-coordinate treatment of the low-frequency vibrations of a hydrogen-bonded crystal of NMA has been performed by Shimanouchi *et al.*,[2, 3] and the polarization vectors have been calculated for $\mathbf{q} = 0$.

In this approximation, the vibrational modes appear as sharp lines at the appropriate frequencies. They are compared in Figure 11.2 with the frequency distribution $G(\omega)$ derived from the neutron spectrum of NMA. The comparison was made in an attempt to predict the relative intensity of the peaks in $G(\omega)$. The broad bands observed in $G(\omega)$ are explained by the fact that frequencies for all values of **q** are contained in this distribution.

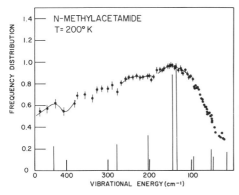

Figure 11.2. Frequency distribution (black dots) of N-methylacetamide at 200°K (solid) derived from the neutron measurements. The vertical lines correspond to calculated values of the eigenvectors for modes involving motions of hydrogen atoms.

11.2   Hexamine

Hexamethylenetetramine (hexamine) is a cubic crystal with two molecules per unit cell. Because of its relative simplicity, it is one of the few molecular crystals for which a model has been proposed,[4, 5] and a normal-coordinate calculation leading to the dispersion curves and the distribution of phonon frequencies has been performed.[4] There are six internal vibration frequencies in hexamine. Becka's model[5] assumes an acoustic branch with a Debye distribution (high-frequency cutoff, 72.6 cm^{-1}) and one triply degenerate optical branch with an Einstein distribution peaked at 44 cm^{-1}, in agreement with Raman and calorimetric measurements.[5]

The frequency distribution $G(\omega)$ derived from the neutron-scattering data is shown in Figure 11.3. It displays two peaks at 39 cm^{-1} and 51 cm^{-1} which are assigned to split optical modes. The peak at 39 cm^{-1} cannot be assigned to a flat acoustic branch, and the high-frequency cutoff is higher than the one used in Becka's model.

Cochran and Pawley[4] have presented a somewhat more refined model of hexamine and have computed the phonon dispersion curves. Assuming the harmonic approximation, they consider interaction only

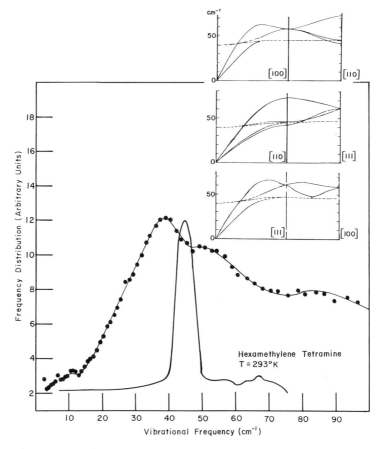

Figure 11.3. The frequency distribution of hexamethylenetetramine at 293°K derived from the neutron measurements (black dots) and the calculated frequency distribution (solid line). The calculated dispersion curves in several directions in the crystal are also shown in the upper part of the figure.

up to the next nearest neighbor and use intermolecular force constants estimated from measured elastic constants and the lowest Raman frequencies. The dispersion curves along three symmetry directions of the crystal and the corresponding frequency distribution $g(\omega)$ are displayed in Figure 11.3 together with $G(\omega)$ derived from the neutron measurements. A comparison of $g(\omega)$ with $G(\omega)$ indicates that the optical modes are split, that they are much broader than predicted by the model, and that the peak corresponding to the translational modes is seen around 86 cm^{-1} in $G(\omega)$. The observed differences in intensity are not too significant in view of the approximations involved in the model.

11.3 Hydrides

Neutrons have been used quite extensively to study the motion and binding of hydrogen atoms in a number of metal hydrides.[6, 7] Incident neutrons with energies up to about 1 eV have been used to observe simultaneous excitation of bound levels up to the fourth level. In zirconium hydride the first three vibrational levels appear to be characterized by an equal spacing of about $\hbar\omega_0 = 1120$ cm^{-1} (Figure 11.4). The angular dependence of the elastic as well as the inelastic parts of the cross section indicates substantial agreement with theoretical predictions based on isotopic and elastic forces between Zr and H atoms.[7] The motion of the H atom, as a first approximation, is repre-

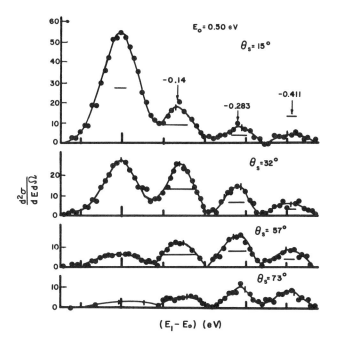

Figure 11.4. Differential scattering cross section for ZrH at scattering angles of 73°, 57°, 32°, and 15°. The energy of the incident neutrons (peak on the left) is $E_0 = 0.504$ eV. The energy transfers in eV are indicated above each peak (0.001 eV = 8 cm^{-1}).

sented by the Einstein model with an isotropic, harmonic oscillator potential. Some of these metal hydrides have a simple cubic-centered structure, and it is possible to use both the neutron and infrared data to obtain the general shape of their dispersion curves.

Saline hydrides of the alkali and alkali earth metals have also been investigated.[8, 9] X-ray and neutron-diffraction data indicate that they have perfect perovskite structures. The vibrational spectra of $BaLiH_3$ and $SrLiH_3$ below 800 cm^{-1} have been obtained (Figures 11.5 and 11.6). A comparison of neutron and infrared frequencies is presented in Table 11.2.[10]

The cubic perovskite structure belongs to the space group O_h^1 and contains five atoms per unit cell. Hunt, Perry, and Ferguson[11] showed by group-theoretical arguments that the twelve nontranslatory modes in the perovskite structure consisted of three triply degenerate normal modes $\gamma_1, \gamma_2, \gamma_3$ belonging to the symmetry species F_{1u} and one triply degenerate mode γ_4 of F_{2u} symmetry. The F_{2u} mode is inactive in the

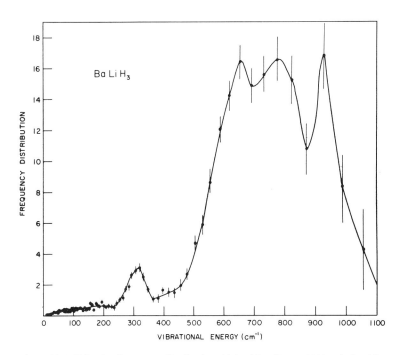

Figure 11.5. Effective frequency distribution $G(\omega)$ of $BaLiH_3$ at 293°K derived from the neutron data.

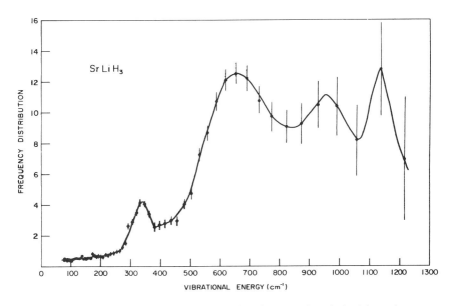

Figure 11.6. Effective frequency distribution of SrLiH$_3$ at 293°K derived from the neutron data.

Table 11.2. Comparison of Vibration Frequencies (cm^{-1}) Observed in Neutron and Infrared Measurements

SrLiH$_3$		BaLiH$_3$	
Neutron Data	IR Data	Neutron Data	IR Data
1130	1080	920	910
950	Inactive	780	Inactive
660	670	650	650
340	—	310	—

infrared. Here γ_1 is the Li—H stretching mode, γ_2 is the infrared-inactive bending mode, γ_3 is the infrared-active H—Li—H bending mode, and γ_4 is the lattice mode in which the M^{++} lattice is displaced relative to a lattice consisting of Li$^+$ and H$^+$ ions. The low-frequency vibration observed in the neutron measurements is identified with γ_4 since it is expected that external oscillations will occur at a much lower frequency than internal ones. This follows from the assumption that forces between groups are much weaker than those among atoms

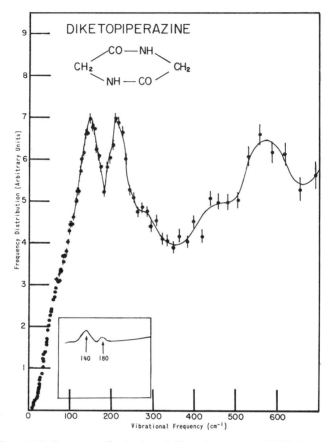

Figure 11.7. Frequency distribution of diketopiperazine at 293°K derived from the neutron measurements. In the insert the far infrared spectrum is shown for comparison.

within a group. The low-frequency vibration was not observed in the infrared because of the KBr cutoff. The high-frequency band is associated with the Li—H stretching mode, γ_1, and is sensitive to the Li—H distance. The Li—H distance is much less in $SrLiH_3$ than in $BaLiH_3$. Consequently the Li—H force constant for $SrLiH_3$ is larger and the corresponding frequency γ_1 increases. The γ_2 frequency is forbidden in the infrared but clearly present in the neutron spectra. However, there appears to be a small shoulder in the infrared spectra of $SrLiH_3$ in the position where γ_2 is expected. This is probably due to relaxation of the

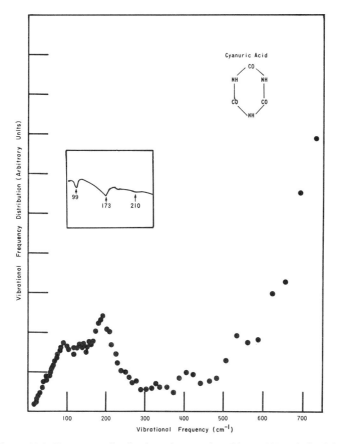

Figure 11.8. Frequency distribution of cyanuric acid at 293°K derived from the neutron measurements. In the insert, the far infrared spectrum is shown for comparison.

selection rules as a result of defects in the perovskite lattice. The last frequency is assigned to the γ_3 bending mode and is observed in the infrared as well as with neutrons.

11.4   Ring Compounds

The spectra of scattered neutrons from polycrystalline samples of diketopiperazine and cyanuric acid are shown in Figures 11.7 and 11.8. The vibration frequencies obtained from neutron[1] and infrared[11]

Table 11.3. Vibration Frequencies (cm^{-1}) in Diketopiperazine

Observed		Calculated[a]	Assignment
Neutron	Infrared		
774	806	728	Ring deformation (in-plane)
570			NH out-of-plane bending
440	449	403	C=O in-plane bending and C—CH$_2$ stretching
285			
215			C—N torsion
168	180		Ring out-of-plane deformation
147	140		Ring out-of-plane deformation
70 (broad)			Lattice modes

[a] T. Shimanouchi and I. Harada, *J. Chem. Phys.* **41**, 2651 (1964).

Table 11.4. Vibration Frequencies (cm^{-1}) in Cyanuric Acid

Observed			Calculated
Raman Data	Infrared Data	Neutron Data	
			15
			29
			39
		56	55
			64
			65
			73
		80	
	99		90
138		128	146
	173	184	
	210		
		369	
		552	
		768	

spectra are assembled in Tables 11.3 and 11.4, together with the results of a partial normal-mode calculation. There is no intermolecular hydrogen bonding in diketopiperazine, so the lattice modes are expected to occur at very low frequencies as is observed in the neutron spectrum (around 70 cm^{-1}). In contrast, strong intermolecular hydrogen bonds give rise to several intense bands in the neutron spectrum of cyanuric acid (Figure 11.8) below 200 cm^{-1}. The intense line at 184 cm^{-1} is resolved into two weak peaks in the far-infrared spectrum at 210 and 173 cm^{-1} (insert in Figure 11.8), which have been assigned[11] to out-of-plane ring deformations. However, because of the intensity of the peak in the neutron spectrum, it is assigned to an amide VII frequency in the peptide group.

REFERENCES

1. H. Boutin, unpublished results.
2. T. Shimanouchi, private communication.
3. T. Shimanouchi and I. Harada, *J. Chem. Phys.* **41**, 2651 (1964).
4. W. Cochran and G. S. Pawley, *Proc. Roy. Soc. (London)* **A280**, 1 (1964).
5. L. N. Becka, *J. Chem. Phys.* **37**, 431 (1962).
6. W. L. Whittemore, in *Inelastic Scattering of Neutrons in Solids and Liquids* (International Atomic Energy Agency, Vienna, 1965), Vol. 2, p. 305.
7. D. K. Harling and B. R. Leonard, Proceedings of the Symposium on Inelastic Scattering of Neutrons by Condensed Systems, Brookhaven National Laboratory, BNL 940 (C–45), p. 96, (1965).
8. C. E. Messer, J. C. Eastman, R. G. Mers, and A. J. Maeland, *Inorg. Chem.* **3**, 776 (1964).
9. C. E. Messer and K. Hardcastle, *Inorg. Chem.* **3**, 1327 (1964).
10. A. S. Maeland, private communication.
11. G. R. Hunt, C. H. Perry, and J. Ferguson, *Phys. Rev.* **A134**, 688 (1964).

12. Experimental Techniques

12.1 Chopper Time-of-Flight Spectrometer

Monochromaticity of incident neutrons is obtained by using a single crystal, a beryllium filter, or a phase rotor system. Energy analysis of the neutrons after scattering can be performed by another single crystal or by a time-of-flight method. A schematic diagram of the experimental facility used for some of the measurements presented in this monograph is shown in Figure 12.1. The thermal-neutron beam from the reactor is filtered by polycrystalline beryllium. The neutron-scattering cross section of beryllium exhibits a sharp drop at a neutron energy of 0.005 eV. Thus, neutrons whose energy is less than 0.005 eV are allowed to pass through the filter and all others are scattered out of the beam. The polycrystalline beryllium filter ($2.5 \times 2.5 \times 16$ in.) is maintained at liquid nitrogen temperatures to increase the transmission of neutrons by decreasing the number of neutrons scattered out of the beam. The filtered beam is not truly monochromatic, but the width of the spectrum is small enough to permit meaningful measurements of the amount of energy gained by the neutron in the scattering. The filtered beam is collimated before impinging upon the sample under investigation. The sample to be studied (generally in polycrystalline form) is placed at the end of the collimator. The energy of the neutrons scattered at different angles is then measured by timing the flight of the neutrons to the detector. This is accomplished by "chopping" the scattered beam with a set of rotating slits. The chopper thus provides a reference time at

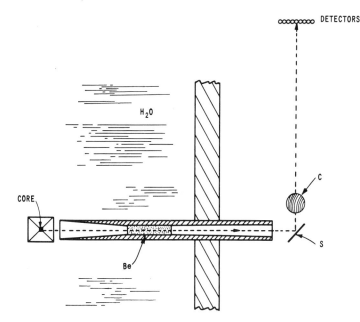

Figure 12.1. Schematic of a cold neutron facility.

which the neutrons can be said to initiate their flight to the detectors. The chopper consists of 32 blades (0.0215-in. thick) and 33 spaces (0.062 in.). Each blade is a sandwich-type structure of Ni (0.001 in.), Cd (0.002 in.), and polyethylene (0.002 in.). The design of the chopper is, of course, quite critical to the resolution of the measurements. The chopper is rotated at speeds varying between 7,000 and 10,000 rpm and is placed at the beginning of a 5-m-long flight path. The flight path is an evacuated tube at the end of which the neutron detectors are placed in an array. These detectors can be filled with enriched boron trifluoride BF_3 (96 percent B^{16}) or with a helium isotope (He^3). The lower part of the flight tube as well as the chopper-collimator-sample assembly are shielded to eliminate the fast-neutron background in the reactor room. An opening in the shield permits rapid access to the sample under study. The flight time of the neutrons from the center of the chopper to the detectors is measured with an electronic clock. This clock consists of 256 channels, each 32 μsec long. Each channel records the arrival of neutrons whose flight times correspond to that particular channel. Thus, a neutron whose flight time is 320 μsec is recorded in channel 10.

In this manner, the entire spectrum of scattered neutrons is simultaneously measured.

The energy resolution is about 2 percent at low energy transfer and becomes worse if larger energy transfers are sought. In general, the instrument can be said to be effective in the observation of modes whose energies range from 40 to 1000 cm^{-1}.

Any improvement in these measurements is strongly dependent upon the intensity of the neutron flux and the resolution of the experimental apparatus. Intensity of neutron flux increases linearly with the power of the reactor. The significant quantity, however, is the signal-to-noise ratio, and the noise (or background of fast neutrons and γ rays) depends upon the type of reactor available. High-flux reactors allow a neutron spectrum to be obtained, with adequate statistics, in about one hour. The intensity of long-wavelength neutrons, which are the ones used in these experiments, can be increased at a given reactor power by the use of a cold moderator (liquid H_2 or liquid CH_4) so as to shift the Maxwellian distribution of neutrons toward lower energy.

There are several ways of improving the resolution of the experimental apparatus, although always at a cost in intensity. Better resolutions can be achieved by getting an incident beam of neutrons as monochromatic as possible (using a single crystal or a phase chopper system) or by increasing the length of the flight path. Alternatively, one can improve the chopper characteristics or decrease the channel width of the multichannel analyzer. Improvements such as these yield a neutron-scattering resolution comparable to that of far-infrared techniques.

12.2   Other Experimental Techniques

Next to the chopper time-of-flight technique, the most useful of the available methods of measurement are the crystal spectrometer methods and the pulsed-monoenergetic-beam/time-of-flight method.* Pulsed-beam spectrometers have the advantage of higher counting rates than crystal spectrometers, especially in view of the fact that simultaneous measurements in several counters at different angles of scattering are possible. The crystal spectrometer permits several valuable new types of specialized experiment to be performed, among

---

* For a more detailed description of these two techniques see B. N. Brockhouse in *Inelastic Scattering of Neutrons in Solids and Liquids* (International Atomic Energy Agency, Vienna, 1961), p. 113.

them energy-distribution measurements at constant momentum transfer (the "constant-Q" method).

12.2.1 Triple-Axis Crystal Spectrometer

The four principal components of the triple-axis crystal spectrometer are (1) the monochromator facility that produces the mono-energetic neutron beam; (2) the positional spectrometer that sets the angles θ and ψ; (3) the analyzing spectrometer that measures the outgoing energy; (4) the control units and recording apparatus.

The monochromator facility comprises the shielding, collimators, and mechanism for the monochromating crystal (X_1); a large movable platform that carries the positional and angular spectrometers; and an accurate scale by which the angle $2\theta_M$ can be read. The facility is designed primarily to be used as a crystal spectrometer, but it can also be used as the basis for a double-chopper time-of-flight spectrometer or for a rotating-crystal spectrometer.

The positional spectrometer consists of a heavy arm whose angular position (ϕ) is indicated directly on an accurate scale, and an accurately indexed table (ψ) that can be coupled to the arm through a half-angling mechanism. The spectrometer arm can be driven electrically (in $\frac{1}{8}^\circ$ steps, for example) and the indexed table arranged to half-angle, or it can be driven independently, also in $\frac{1}{8}^\circ$ steps.

The analyzing spectrometer is mounted directly on the arm of the positional spectrometer. The analyzing crystal is mounted on an indexed table (providing the angle $2\theta_A$) permanently connected to the arm of the analyzing spectrometer through a half-angling mechanism. The arm is driven electrically (in $\frac{1}{8}^\circ$ steps, for example) at an angular speed of several degrees/minute by means of two worm-gear assemblies. There is no mechanism for decoupling the spectrometer from the worm-gear drive, and all movements are made electrically. The motor, mounted at the end of the spectrometer arm, is connected to the worm-gear drive through a magnetic clutch. The magnetic clutch is activated for the brief time necessary to go through $\frac{1}{8}^\circ$ cycles.

The control units provide the following sequence of operations:

1. The spectrometers are set up manually with the four parameters $2\theta_M$, $2\theta_A$, ϕ, and ψ at their initial positions. At least one of the parameters is fixed. The numbers of $\frac{1}{8}^\circ$ increments by which the other (moving) parameters will advance are selected.

2. The spectrometers advance to their first positions.

3. Counting takes place at the first position. The number of counts accumulated in the signal counter for a preset number of monitor counts is recorded. The counting period may be repeated without change, it may be repeated with insertion of a cadmium shutter to measure background, or it may be repeated with the analyzing crystal turned out of the Bragg position for a different kind of background measurement.

4. After completion of the operations at the first positions of the parameters, the spectrometers again advance by the preselected increments, and counting is repeated. This process continues through a preset number of positions. At the conclusion of a set of measurements (for some types of experiments), the spectrometers can be made to move to new initial positions, and a new set of measurements can be carried out automatically.

12.2.2 *The Rotating-Crystal Time-of-Flight Spectrometer*

The rotating-crystal spectrometer belongs to the class of *pulsed-beam* spectrometers in which a pulsed monoenergetic beam of neutrons is scattered by a specimen, and the energy distributions of the neutrons scattered through various angles are measured by time-of-flight. In general, pulsed-beam spectrometers are capable of higher resolution for a given intensity than are crystal spectrometers. (However, they do not permit use of the "constant-Q" method or other special methods.) The rotating-crystal instrument is the simplest and cheapest of the pulsed-beam spectrometers, but it has a number of peculiarities that operate both to its advantage and to its disadvantage. A single crystal rotates about a vertical shaft at a comparatively high angular velocity (about 8000 rpm). Each time a crystal plane comes into position for Bragg reflection, a burst of monoenergetic neutrons, with energy given by the Bragg law, is sent down the collimator and impinges on the specimen. The scattered neutrons are observed by means of several counters at different angles of scattering, and their energies after scattering are determined from their times-of-flight over the known distances from the specimen to the counters.

The method has several attractive features:

1. The equipment is comparatively simple and cheap, and it is rugged and reliable.

2. Backgrounds are low since the monoenergetic beam is deflected away from the line of the experimental hole.

3. There is no time-dependent fast-neutron background.

4. The pulsed monoenergetic beam can have a larger area than is readily obtainable with beams produced by mechanical means.

5. As in other pulsed-beam time-of-flight methods, measurements may be made simultaneously at several angles of scattering, thus enhancing the counting rate.

6. Burst-focusing properties sometimes enable long on-time to be used without spoiling the resolution.

As in crystal spectrometers, order contamination exists in the monoenergetic beam. If the rotating crystal produces only two bursts per revolution (that is, if a single plane is used), it can be arranged that the only contaminant pulses will be those from the higher orders of the plane. If, however, four or six bursts per revolution are produced, other planes necessarily come into reflecting position and contaminate the beam.

In addition to order contamination, the beam contains neutrons diffusely scattered by the rotating crystal and its shield. These neutrons have a distribution of energies and are scattered continuously in time, so they produce a flat background which is almost time-independent. This "white" contamination is much more important with a rotating-crystal than with an ordinary crystal spectrometer because the Bragg-scattered neutrons are reflected only for a small fraction of the time (perhaps 1 percent) while the diffusely scattered neutrons are produced all the time. It is therefore very important that a crystal for use in a rotating-crystal spectrometer have little incoherent scattering. It should also have a melting temperature much greater than room temperature so that thermal diffuse scattering can be kept small. (A high melting point is also desirable from the point of view of mechanical strength.)

All order contamination can be removed and the white contamination reduced by passing either the incident or the diffracted beam through a coarse chopper phased with the rotating crystal so as to pass first-order neutrons. At some cost in intensity but with an improvement in resolution, this can also be accomplished by reflecting the beam again from another rotating crystal phased with the first. Filters may also be used, as in normal crystal spectrometers. At present the instrument is being commonly used with neutrons of wavelength longer than the beryllium cutoff, and a liquid-nitrogen-cooled beryllium filter eliminates the order of contamination.

In common with all pulsed-beam spectrometers the instrument

suffers from frame overlap. It is not clear beforehand with what initial neutron burst a detected neutron is to be identified. Hence the time-of-flight of a neutron is ambiguous to the extent that any integral number of repetition times may be added to the nominal time-of-flight. Frame overlap for a given spectrum can be almost eliminated by reducing the repetition rate of the monoenergetic pulses, but usually only at a considerable cost in intensity.

Especially at long wavelengths, the operation of the instrument is influenced by the Doppler shifts in the neutron velocities produced by the crystal rotation. These are correlated with the positions in the crystal at which reflection takes place since different parts of the crystal have different velocities. The shifts can sometimes be used to compensate for the uncertainty in the time of origin of the burst, which is introduced by the dimensions of the rotating crystal. Furthermore, as the crystal rotates, the length of the path followed by the neutron may gradually change, as may also the reflected wavelength. Thus there is the possibility of *burst-focusing*, since neutrons reflected at different times with different wavelengths in different parts of the crystal can be made to arrive at the counter simultaneously.

Because planes of high multiplicity can be used (up to six in a cubic crystal), the repetition rate of the pulses can be high with quite moderate speeds of rotation.

The rotating-crystal spectrometer is inherently a high-resolution instrument since, for it to be effective, the collimation must approximately match the effective mosaic width of the crystal. This width is usually not large, nor can a crystal of very large mosaic width be profitably used even if available because of the resulting low peak reflectivity. Thus, the angular apertures are kept small, the burst time short, and the resolution high.

A. Appendix

Derivation of Double Differential Scattering Cross Section

In this appendix we will derive the basic expression for the double differential neutron-scattering cross section. We will use a slightly different approach from that given in Section 2.1 but will arrive at the same results. For the purpose of the derivation it is not necessary to specify the physical properties of the scattering system; consequently the results are applicable to any measurement of neutron-nuclear scattering at thermal or subthermal energies.

Consider the idealized experiment shown in Figure A.1. An incident plane wave of monoenergetic neutrons with energy E_i and wave vector $\mathbf{k}_i$ is scattered, and the direction of scattering θ and final neutron energy E_f are measured. The angular differential cross section for this process is

$$\frac{d\sigma}{d\Omega} = \frac{R_d^2 \hat{\mathbf{r}} \cdot \mathbf{J}_f(R_d)}{\hat{\mathbf{Z}} \cdot \mathbf{J}_i}, \qquad (A.1)$$

where $\mathbf{J}$ is the neutron current and subscripts i and f denote initial and final conditions. The detector is located a distance R_d away, which is large compared to the target dimensions. The incident neutron wave function is a plane wave, $\exp(i\mathbf{k}_i \cdot \mathbf{r})$, and since

$$\mathbf{J} = \frac{\hbar}{2m} \left[\Psi^*(\nabla\Psi) - (\nabla\Psi^*)\Psi \right], \qquad (A.2)$$

where m is neutron mass and Ψ the wave function, we have $\hat{\mathbf{Z}} \cdot \mathbf{J}_i = \hbar k_i/m = v_i$. To calculate the scattered-neutron wave function or $\mathbf{J}_f$ we employ the stationary Schrödinger equation

211

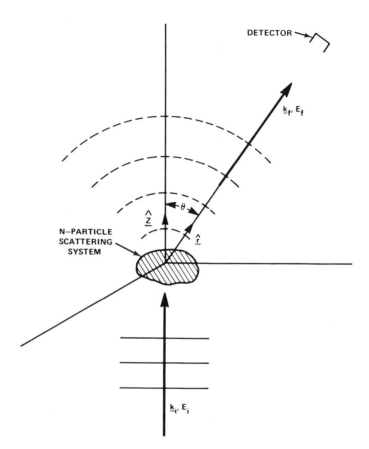

Figure A.1. Schematic diagram of neutron scattering experiment.

$$\left(\frac{p^2}{2m} + H_s + U\right)\Phi(\mathbf{r}, \mathbf{R}^N) = \lambda\Phi(\mathbf{r}, \mathbf{R}^N), \qquad (A.3)$$

where $p^2/2m$ is the kinetic energy operator of the neutron, H_s the Hamiltonian of the scattering system, $U(\mathbf{r}, \mathbf{R}^N)$ the neutron-nuclear interaction, and λ the total energy. Here $\mathbf{r}$ denotes neutron position and $\mathbf{R}^N = \{\mathbf{R}_1, \mathbf{R}_2, ..., \mathbf{R}_N\}$ denotes the positions of N nuclei in the target. Let the eigenstates and eigenvalues of the scattering system be $|n\rangle$ and ϵ_n; then

$$H_s|n\rangle = \epsilon_n|n\rangle, \qquad (A.4)$$

$$\langle n|n'\rangle = \delta_{nn'},$$

and for simplicity we assume that the eigenvalue is discrete and that the system is nondegenerate. One can expand as follows:

$$\Phi(\mathbf{r}, \mathbf{R}^N) = \sum_n \Psi_n(\mathbf{r}) |n\rangle. \tag{A.5}$$

Then Equation A.3 becomes

$$(\nabla^2 + k_n^2) \Psi_n(\mathbf{r}) = \frac{2m}{\hbar^2} \sum_{n'} \langle n|U|n'\rangle \Psi_{n'}(\mathbf{r}) \tag{A.6}$$

with

$$k_n^2 = \frac{2m}{\hbar^2} (\lambda - \epsilon_n). \tag{A.7}$$

To solve Equation A.6 we make use of the free-particle Green's function

$$G_n(|\mathbf{r} - \mathbf{r}'|) = \frac{\exp(ik_n|\mathbf{r} - \mathbf{r}'|)}{4\pi|\mathbf{r} - \mathbf{r}'|}, \tag{A.8}$$

which is the solution to

$$(\nabla^2 + k_n^2) G_n(|\mathbf{r} - \mathbf{r}'|) = -\delta(\mathbf{r} - \mathbf{r}'). \tag{A.9}$$

Equation A.6 can therefore be converted into an integral equation,

$$\Psi_n(\mathbf{r}) = \Psi_n^0(\mathbf{r}) - \frac{2m}{\hbar^2} \sum_{n'} \int d^3 r' \, G_n(|\mathbf{r} - \mathbf{r}'|) \langle n|U|n'\rangle \Psi_{n'}(\mathbf{r}'), \tag{A.10}$$

where

$$\Psi_n^0(\mathbf{r}) = \delta_{nn_0} \exp(i\mathbf{k}_i \cdot \mathbf{r}) \tag{A.11}$$

satisfies $(\nabla^2 + k_n^2) \Psi_n^0(\mathbf{r}) = 0$ and where the initial system state is taken to be $|n_0\rangle$. Since the r' integral is restricted to the target volume at the detector, we have $r/r' \gg 1$ and thus asymptotically

$$\frac{\exp(ik_n|\mathbf{r} - \mathbf{r}'|)}{|\mathbf{r} - \mathbf{r}'|} \simeq \frac{\exp(ik_n r)}{r} \exp(i\mathbf{k}_n \cdot \mathbf{r}'). \tag{A.12}$$

Inserting Equation A.12 into Equation A.10 and iterating once (the first Born approximation) with the inhomogeneous term, we obtain

$$\Psi_n(\mathbf{r}) \simeq \delta_{nn_0} \exp(i\mathbf{k}_i \cdot \mathbf{r}) - \frac{\exp(ik_n r)}{r} \left(\frac{m}{2\pi\hbar^2}\right) \int d^3 r' \exp\left[i(\mathbf{k}_i - \mathbf{k}_n) \cdot \mathbf{r}'\right]$$
$$\times \langle n|U|n_0\rangle. \tag{A.13}$$

From this we see that the scattering amplitude for a transition from system state $|n_0\rangle$ to $|n\rangle$ is

$$f_{nn_0} = -\frac{m}{2\pi\hbar^2}\int d^3r' \exp\left[i(\mathbf{k}_i - \mathbf{k}_n)\cdot\mathbf{r}'\right] \langle n|U|n_0\rangle. \qquad (A.14)$$

The corresponding angular differential cross section per nucleus becomes

$$\frac{d\sigma_{nn_0}}{d\Omega} = \frac{k_n}{k_i N}|f_{nn_0}|^2. \qquad (A.15)$$

In practice, the initial state of the target cannot be prepared nor can we observe the final state; consequently the measured quantity is an average of Equation A.15 over all initial states and a sum over all final states. Denoting the probability that the target system is initially in state n by $P(n)$, we have

$$\frac{d\sigma}{d\Omega} = N^{-1}\sum_{n_0 n} P(n_0)\frac{k_n}{k_i}|f_{nn_0}|^2. \qquad (A.16)$$

We obtain the energy and angular differential cross section by first defining

$$\frac{d\sigma}{d\Omega} = \int dE_f\,\frac{d^2\sigma}{d\Omega dE} \qquad (A.17)$$

with

$$\frac{d^2\sigma}{d\Omega dE} = N^{-1}\sum_{nn_0} P(n_0)\frac{k_n}{k_i}|f_{nn_0}|^2\,\delta(E_f + \epsilon_n - E_i - \epsilon_{n_0}) \qquad (A.18)$$

and $E_f = \hbar k_n^2/2m$. It is conventional to take for $U(\mathbf{r}, \mathbf{R}^N)$ the so-called Fermi pseudopotential,

$$U(\mathbf{r}, \mathbf{R}^N) = \frac{2\pi\hbar^2}{m}\sum_{l=1}^{N} a_l\,\delta(\mathbf{r} - \mathbf{R}_l), \qquad (A.19)$$

where a_l is the bound-nucleus scattering length of the lth nucleus. This potential describes the collision as a localized impact and corresponds to hard-sphere interaction at low energies. The use of the first Born approximation in conjunction with Equation A.19 leads to the correct free-atom scattering result. The approximation is considered to introduce no significant error in cross section calculations. Inserting Equation A.19 into Equation A.14 we find*

$$\frac{d^2\sigma}{d\Omega dE} = \left(\frac{E_f}{E_i}\right)^{1/2}\sum_{nn_0} P(n_0)\,\delta(\epsilon_n - \epsilon_{n_0} + \hbar\omega)\frac{1}{N}|\langle n_0|\sum_l a_l \exp\left(+i\boldsymbol{\kappa}\cdot\mathbf{R}_l\right)|n\rangle|^2,$$
$$(A.20)$$

* Notice that because of the delta function we can take k_n out of the sum and denote it as k_f.

where $\hbar\omega = E_f - E_i$ and $\boldsymbol{\kappa} = \mathbf{k}_i - \mathbf{k}_f$. This is the result quoted in Equation 2.4.

The sum over the final states $|n\rangle$ in Equation A.20 can be carried out using the closure property. The integral representation of the delta function

$$\delta(x) = \frac{1}{2\pi} \int_{-\infty}^{\infty} dt \, \exp\left(-itx\right) \tag{A.21}$$

enables us to write

$$\sum_n \delta(\epsilon_n - \epsilon_{n_0} + \hbar\omega) \langle n_0 | \exp\left(i\boldsymbol{\kappa}\cdot\mathbf{R}_l\right) | n \rangle \langle n | \exp\left(-i\boldsymbol{\kappa}\cdot\mathbf{R}_{l'}\right) | n_0 \rangle$$

$$= \frac{1}{2\pi\hbar} \int_{-\infty}^{\infty} dt \, \exp\left(-i\omega t\right) \sum_n \langle n_0 | \exp\left(itH_s/\hbar\right) \exp\left(i\boldsymbol{\kappa}\cdot\mathbf{R}_l\right) \exp\left(-itH_s/\hbar\right) | n \rangle$$

$$\times \langle n | \exp\left(-i\boldsymbol{\kappa}\cdot\mathbf{R}_{l'}\right) | n_0 \rangle$$

$$= \frac{1}{2\pi\hbar} \int_{-\infty}^{\infty} dt \, \exp\left(-i\omega t\right) \langle n_0 | \exp\left[i\boldsymbol{\kappa}\cdot\mathbf{R}_l(t)\right] \exp\left(-i\boldsymbol{\kappa}\cdot\mathbf{R}_{l'}\right) | n_0 \rangle, \tag{A.22}$$

where $R(t)$ is the Heisenberg position operator,

$$\frac{d\mathbf{R}}{dt} = \frac{\hbar}{i}\left[\mathbf{R}(t), H_s\right]. \tag{A.23}$$

(When written without an argument, $\mathbf{R}$ is understood to mean at $t = 0$.) Thus, Equation A.20 becomes*

$$\frac{d^2\sigma}{d\Omega dE} = \left(\frac{E_f}{E_i}\right)^{1/2} \frac{1}{2\pi\hbar N} \int_{-\infty}^{\infty} dt \, \exp\left(-i\omega t\right)$$

$$\times \sum_{ll'} a_l a_{l'} \langle \exp\left[i\boldsymbol{\kappa}\cdot\mathbf{R}_l(t)\right] \exp\left(-i\boldsymbol{\kappa}\cdot\mathbf{R}_{l'}\right)\rangle, \tag{A.24}$$

where $\langle\,\rangle$ is defined by Equation 2.22. Strictly speaking, Equation A.24 is correct for a system of nuclei with zero spin; otherwise appropriate averages of the scattering length would have to be taken. The results in this case are discussed in Section 2.1. Equation A.24, which is essentially Equation 2.19 without the separation into coherent and incoherent contributions, is the basic starting point for the analysis of inelastic-neutron-scattering experiments. Its compact appearance tends to mask the complicated ways in which structural and dynamic properties of the medium can influence the cross section.

---

* It will be sometimes convenient to write $d^2\sigma/d\Omega d\omega = \hbar d^2\sigma/d\Omega dE$.

B. Appendix

Scattering by an Oscillator

In this appendix we will describe the calculation of the incoherent differential scattering cross section of a simple harmonic oscillator (mass M, frequency ω) in a medium at temperature T.* We begin with the intermediate scattering function, Equation 2.25:

$$\chi_s(\kappa, t) = \langle \exp[i\kappa X(t)] \exp(-i\kappa X) \rangle. \tag{B.1}$$

Since we consider only isotopic vibrations, it is sufficient to treat the one-dimensional problem. Generalization of our results to three dimensions is obvious. The oscillator position at time t can be written as

$$X(t) = \left(\frac{\hbar}{2M\omega}\right)^{1/2} [a^+ \exp(i\omega t) + a \exp(-i\omega t)], \tag{B.2}$$

where a^+ and a are called "creation" and "destruction" operators because of the properties

$$\langle n+1|a^+|n \rangle = \sqrt{n+1},$$
$$\langle n-1|a|n \rangle = \sqrt{n}, \tag{B.3}$$

where $|n\rangle$ is the nth eigenstate of the oscillator. Equation B.2 is a convenient representation of $X(t)$ and is of course another statement of the result

$$X(t) = X \cos \omega t + \frac{P}{M\omega} \sin \omega t, \tag{B.4}$$

where X and P are the position and momentum at $t = 0$.

---

* See, for example, A. Messiah, *Quantum Mechanics* (John Wiley & Sons, Inc., New York, 1966), Vol. I, Chap. 12, for discussions of oscillator properties.

To evaluate Equation B.1 we make use of the fact that while $X(t)$ and X do not commute, their commutator is simply a c number,

$$[X(t), X] = \frac{\hbar}{2M\omega} [\exp(-i\omega t) - \exp(i\omega t)]. \qquad (B.5)$$

Equation B.5 follows directly from the commutation relation between X and P or from

$$[a, a^+] = 1, \qquad (B.6)$$
$$[a, a] = [a^+, a^+] = 0.$$

Thus, we can apply the identity

$$\exp(A)\exp(B) = \exp(A + B + \tfrac{1}{2}[A, B]), \qquad (B.7)$$

which holds for any two operators A and B provided that they both commute with $[A, B]$. Equation B.1 now becomes

$$\chi_s(\kappa, t) = \exp\{\tfrac{1}{2}\kappa^2 [X(t), X]\}\langle\exp\{i\kappa[X(t) - X]\}\rangle. \qquad (B.8)$$

It can be further reduced through a theorem stating that if Q is any multiple or linear combination of commuting oscillator coordinates and their conjugate momenta,*

$$\langle\exp Q\rangle = \exp(\tfrac{1}{2}\langle Q^2\rangle). \qquad (B.9)$$

Then

$$\chi_s(\kappa, t) = \exp\{\tfrac{1}{2}\kappa^2 [X(t), X]\}\exp\{-\tfrac{1}{2}\kappa^2\langle[X(t) - X]^2\rangle\}$$
$$= \exp\{-\kappa^2[\langle X^2\rangle - \langle X(t)X\rangle]\} \qquad (B.10)$$

in view of the fact that $\langle X^2(t)\rangle = \langle X^2\rangle$. The evaluation of $\langle X^2\rangle$ and $\langle X(t)X\rangle$ can be quite simply carried out in terms of the operators a^+ and a. By $\langle X^2\rangle$ we mean

$$\langle X^2\rangle = \sum_u P(n)\langle n|X^2|n\rangle, \qquad (B.11)$$

where $P(n)$ is the probability that the oscillator is in state n,

$$P(n) = \exp(-\epsilon_n/k_bT)\left[\sum_{n'}\exp(-\epsilon_{n'}/k_bT)\right]^{-1}. \qquad (B.12)$$

Since $\epsilon_n = \hbar\omega(n + \tfrac{1}{2})$,

$$P(n) = \exp(-2nz)[1 - \exp(-2z)], \qquad (B.13)$$

* A. Messiah, *Quantum Mechanics*; also A. C. Zemach and R. J. Glauber, *Phys. Rev.* **101**, 118 (1956).

where $z = \hbar\omega/2k_bT$. Combining Equations B.2, B.3, and B.11 we find

$$\langle X^2 \rangle = \frac{\hbar}{2M\omega}[\langle a^+ a \rangle + \langle a\, a^+ \rangle]$$

$$= \frac{\hbar}{2M\omega}[2\langle n \rangle + 1] \tag{B.14}$$

and

$$\langle n \rangle = \sum_u n\, P(n)$$

$$= \exp(-2z)[1 - \exp(-2z)]. \tag{B.15}$$

Hence,

$$\langle X^2 \rangle = \frac{\hbar}{2M\omega} \coth z. \tag{B.16}$$

In a similar way we can show that

$$\langle X(t)X \rangle = \frac{\hbar}{2M\omega}[(\langle n \rangle + 1)\exp(-i\omega t) + \langle n \rangle \exp(i\omega t)]$$

$$= \frac{\hbar}{2M\omega}[1 - \exp(-2z)]^{-1}\; [\exp(-i\omega t) + \exp(-2z + i\omega t)]. \tag{B.17}$$

Putting Equations B.16 and B.17 into Equation B.10, we obtain

$$\chi_s(\kappa, t) = \exp\left[-\tfrac{1}{2}\kappa^2\gamma(t)\right], \tag{B.18}$$

$$\gamma(t) = \frac{\hbar}{M\omega}[\coth z(1 - \cos \omega t) + i \sin \omega t], \tag{B.19}$$

which are the results quoted in Equations 2.42 and 2.43. To show that the incoherent scattering law can be written as in Equation 2.44, we rewrite Equation B.18 as

$$\chi_s(\kappa, t) = \exp(-2W)\exp\left[\tfrac{1}{2}\eta(\alpha + 1/\alpha)\right], \tag{B.20}$$

where

$$2W = \kappa^2\langle X^2 \rangle. \tag{B.21}$$

$$\eta = \frac{\hbar\kappa^2}{2M\omega} \operatorname{csch} z, \tag{B.22}$$

$$\alpha = \exp(-z + i\omega t). \tag{B.23}$$

The second exponential in Equation B.20 is in the form of a generating function of the modified Bessel function

$$\exp\left[\tfrac{1}{2}\eta(\alpha+1/\alpha)\right] = \sum_{n=-\infty}^{\infty} \alpha^n I_n(\eta). \qquad \text{(B.24)}$$

Thus

$$\chi_s(\kappa, t) = \exp(-2W) \sum_{n=-\infty}^{\infty} \exp\left(-\frac{n\hbar\omega}{2k_bT}\right)\exp(in\omega t)$$

$$I_n\left(\frac{\hbar\kappa^2}{2M\omega}\operatorname{csch}\frac{\hbar\omega}{2k_bT}\right). \qquad \text{(B.25)}$$

The Fourier transform of this result gives Equation 2.44.

Index